LECTURES ON DIFFERENTIAL EQUATIONS

By

SOLOMON LEFSCHETZ

PRINCETON UNIVERSITY PRESS
Princeton, New Jersey

Published by Princeton University Press
41 William Street
Princeton, New Jersey 08540

Library of Congress Card No. 1578
ISBN 0-691-08395-9

Princeton University Press books are printed on acid-free paper and
meet the guidelines for permanence and durability of the
Committee on Production Guidelines for Book
Longevity of the Council on Library Resources

Printed in the United States of America

PREFACE

The subject of differential equations in the large would seem to offer a most attractive field for further study and research. Many hold the opinion that the classical contributions of Poincaré, Liapounoff and Birkhoff have exhausted the possibilities. This is certainly not the opinion of a large school of Soviet physico mathematicians as the reader will find by consulting N. Minorsky's recent Report on Non-Linear Mechanics issued by the David Taylor Model Basin. In recent lectures at Princeton and Mexico, the author endeavored to provide the necessary background and preparation. The material of these lectures is now offered in the present monograph.

The first three chapters are self-explanatory and deal with more familiar questions. In the presentation vectors and matrices are used to the fullest extent. The fourth chapter contains a rather full treatment of the asymptotic behavior and stability of the solutions near critical points. The method here is entirely inspired by Liapounoff, whose work is less well known that it should be. In Chapter V there will be found the Poincaré-Bendixson theory of planar characteristics in the large. The very short last chapter contains an analytical treatment of certain non-linear differential equations of the second order, dealt with notably by Liénard and van der Pol, and of great importance in certain applications.

The author wishes to express his indebtedness to Messrs. Richard Bellman and Jaime Lifshitz for many valuable suggestions and corrections to this monograph. The responsibility, however, for whatever is still required along that line is wholly the author's.

TABLE OF CONTENTS

CHAPTER I

SOME PRELIMINARY QUESTIONS

§1. MATRICES

1. The reader is assumed familiar with the elements of matrix theory. The matrices $\| a_{ij} \|$, $\| x_{ij} \|$, ..., are written A, X, ... The transpose of A is written A'. The matrix diag $(A_1, \ldots, A_r)$ is

$$\begin{Vmatrix} A_1, & 0, & \cdots & \\ 0, & A_2, & & \\ & & \ddots & A_r \end{Vmatrix}$$

where the A_i are square matrices and the zeros stand for zero matrices. Noteworthy special case: $\mathrm{diag}(a_1, \ldots, a_n)$ denotes a square matrix of order n with the scalars a_i down the main diagonal and the other terms zero. In particular if $a_1 = \ldots = a_n = 1$, the matrix is written E_n or E and called a unit-matrix. The terms of E_n are written δ_{ij} and called Kronecker deltas.

2. Suppose now A square and of order n. The determinant of A is denoted by $|A|$. When $|A| = 0$, A is said to be <u>singular</u>. A non-singular matrix A possesses an inverse A^{-1} which satisfies $AA^{-1} = A^{-1}A = E$. The trace of a square matrix A written tr A, is the expression $\sum a_{ii}$. If $A^n = 0$, A is called <u>nilpotent</u>. We recall the relations

$$(AB)^{-1} = B^{-1}A^{-1} \qquad (A^{-1})' = (A')^{-1},$$

where A, B are non-singular.

If $f(\lambda) = a_0 + a_1\lambda + \ldots + a_r\lambda^r$ then $a_0E + a_1A + \ldots + a_rA^r$ has a unique meaning and is written f(A). The polynomial $\phi(\lambda) = |A-\lambda E|$ is known as the <u>characteristic</u> polynomial of A, and its roots as the characteristic roots of A. (See Theorem (3.5) below.)

3. (3.1) Two real [complex] square matrices A, B of same order n are called similar in the real [complex] domain if there can be found a non-singular square real [complex] matrix P of order n such that $B = PAP^{-1}$. This relation is clearly an equivalence. For if we denote it by $\sim$ then the relation is

symmetric: $A \sim B \longrightarrow B \sim A$, since

$$B = PAP^{-1} \Longrightarrow A = P^{-1}BP$$

reflexive: $A \sim A$, since $A = EAE^{-1}$,

transitive: $A \sim B,\ B \sim C \longrightarrow A \sim C$. For if $A = PBP^{-1}$, $B = QCQ^{-1}$ then $A = PQCQ^{-1}P^{-1} = (PQ)C(PQ)^{-1}$.

(3.2) If $A \sim B$ and $f(\lambda)$ is any polynomial then $f(A) \sim f(B)$. Hence $f(A) = 0 \longrightarrow f(B) = 0$.

For if $B = PAP^{-1}$ then $B^r = PA^rP^{-1}$, $kB = P(kA)P^{-1}$, and $P(A_1 + A_2)P^{-1} = PA_1P^{-1} + PA_2P^{-1}$.

(3.3) Similar matrices have the same characteristic polynomial.

For $B = PAP^{-1} \longrightarrow B - \lambda E = P(A-\lambda E)P^{-1}$, and therefore also $|B-\lambda E| = |A-\lambda E|$.

Since the characteristic polynomials are the same, their coefficients are also the same. Only two are of interest: the determinants, manifestly equal, and the traces. If $\lambda_1, \ldots, \lambda_n$ are the characteristic roots then a ready calculation yields

$$\sum \lambda_i = \sum a_{ii} = \operatorname{tr} A.$$

Therefore

(3.4) Similar matrices have equal traces.

For the proof of the following two classical theorems the reader is referred to the standard treatises on the subject:

(3.5) Theorem. If $\phi(\lambda)$ is the characteristic polynomial of A, then $\phi(A) = 0$.

(3.6) Fundamental Theorem. Every complex square matrix A is similar in the complex domain to a matrix of the form diag $(A_1, \ldots, A_r)$ where A_i is of the form

$$(3.6.1)\qquad A_i = \left\| \begin{matrix} \lambda_j, & 0, & 0, & \ldots\ldots, & 0 \\ 1, & \lambda_j, & 0, & \ldots\ldots & 0 \\ 0, & 1, & \lambda_j, & \ldots\ldots & 0 \\ \ldots & \ldots & \ldots & \ldots & \ldots \\ \ldots & \ldots & \ldots & \ldots & \ldots \\ 0, & \ldots & \ldots & 1, & \lambda_j \end{matrix} \right\|$$

with λ_j one of the characteristic roots. There is at least one A_i for each λ_j and if λ_j is a simple root then there is only one $A_i = \| \lambda_j \|$. Hence if the characteristic roots λ_j are all distinct, A is similar to diag $(\lambda_1, \ldots, \lambda_n)$.

By way of illustration when $n = 2$ and $\lambda_1 = \lambda_2$, we have the two distinct types

$$\left\| \begin{matrix} \lambda, & 0 \\ 0, & \lambda \end{matrix} \right\|, \qquad \left\| \begin{matrix} \lambda, & 0 \\ 1, & \lambda \end{matrix} \right\|.$$

(3.7) Real Matrices. When A is real the λ_i occur in conjugate pairs $\lambda_j, \bar{\lambda}_j$ and hence the matrices A_i occur likewise in conjugate pairs $A_j, \bar{A}_j$ where $\bar{A}_j$ is like A_j with $\bar{\lambda}_j$ instead of λ_j. Thus they may be disposed into a sequence $A_1, \ldots, A_k, \bar{A}_1, \ldots, \bar{A}_k, A_{2k+1}, \ldots, A_s$ where the A_{2k+i} correspond to the real λ_j. We will then say that the canonical form is real.

4. Limits, Series. (4.1) Let $\{A_p\}$, $A_p = \| a^p_{ij} \|$ be a sequence of matrices of order n such that $a_{ij} = \lim \{a^p_{ij}\}$ exists for every pair i, j. We then apply the customary "limit" terminology to the sequence $\{A_p\}$ and call $A = \| a_{ij} \|$ its limit. As a consequence we will naturally say that the infinite series $\sum A_p$ is convergent if the n^2 series

$$(4.2)\qquad a_{ij} = \sum a^p_{ij}$$

are convergent and the sum of the series is by definition the matrix $A = \| a_{ij} \|$.

If the a_{ij}^p are functions of a parameter t and the n series $\sum a_{ij}^p$ are uniformly convergent as to t over a certain range then $\sum A_p$ is said to be <u>uniformly convergent</u> as to t over the same range.

(4.3) Let us apply to the A_p's the simultaneous operation $B_p = PA_pQ$ where P, Q are fixed. If we set

$$S_{ij}^{mr} = \sum_{p=m+1}^{r} a_{ij}^p$$

then clearly the corresponding T_{ij}^{mr} for the B's is related to the S_{ij}^{mr} by

$$T_{ij}^{mr} = \sum p_{ih} S_{hk}^{mr} q_{kj}.$$

Now a n.a.s.c. of convergency of (4.2) may be phrased thus: for every $\varepsilon > 0$ there is an N such that $m > N \Longrightarrow |S_{ij}^{mr}| < \varepsilon$ whatever r. If $\alpha = \sup |p_{ih}, q_{kj}|$ then

$$\sum_{i,j} |T_{ij}^{mr}| < n^2\alpha^2 \sum_{i,j} |S_{ij}^{mr}| .$$

Hence the convergence of (4.2) implies the convergence of

$$\sum B_p = \sum PA_pQ$$

whose limit is clearly B = PAQ. In particular

(4.4) *If $\{A_p\}$ converges to A and if $B_p = PA_pP^{-1}$ then $\{B_p\}$ converges to $B = PAP^{-1}$.*

5. Consider a power series with complex coefficients

(5.1) $$f(z) = a_0 + a_1 z + a_2 z^2 + \dots$$

whose radius of convergence $\rho > 0$. If

(5.2) $$X = \| x_{ij} \|$$

is a square matrix of order n we may form the series

$$a_0E + a_1X + a_2X^2 + \dots \tag{5.3}$$

and if it converges its limit will be written $f(X)$.

(5.4) Suppose that $X = \text{diag}\,(X_1, X_2)$. If $g(z)$ is a scalar polynomial then $g(X) = \text{diag}\,(g(X_1), g(X_2))$. Hence in this case (5.3) converges when and only when the same series in the X_i converge and its limit is then $f(X) = \text{diag}\,(f(X_1), f(X_2))$.

(5.5) Theorem. Sufficient conditions for the convergence of (5.3) are that X is nilpotent or else that its characteristic roots are all less than ρ in absolute value.

Whatever the radius of convergence, when X is nilpotent the series is finite and hence evidently convergent. In the general case, X is similar to a matrix of the type described in (3.6). Remembering (5.4) we only need to consider the type (3.6.1). In other words, we may assume that the matrix is (3.6.1) itself. Thus λ is its sole characteristic root and so we will merely have to prove

(5.6) If X is (3.6.1) and $|\lambda| < \rho$ then (5.3) is convergent.

Let us set

$$Z = \left\| \begin{matrix} 0 & & & & \\ 1 & 0 & & & \\ 0 & 1 & \ddots & & \\ & & \ddots & 1 & 0 \end{matrix} \right\| \tag{5.6.1}$$

We verify by direct multiplication that Z^r is obtained by moving the diagonal of units so that it starts at the term in the $(r+1)^{st}$ row and first column (the term $z_{r+1,1}$). Hence

$$Z^n = 0. \tag{5.6.2}$$

Now $X = \lambda E + Z$, and since E commutes with every matrix:

$$X^p = \lambda^p E + \frac{p}{1!}\lambda^{p-1}Z + \dots + \binom{p}{n-1}\lambda^{p-n+1}Z^{n-1} \tag{5.6.3}$$

Hence

(5.7)

$$f(X) = \begin{vmatrix} f(\lambda), & 0, & \dots & & 0 \\ \frac{f'(\lambda)}{1!}, & f(\lambda), & 0 & \dots & 0 \\ \frac{f''(\lambda)}{2!}, & \frac{f'(\lambda)}{1!}, & f(\lambda), 0, & \dots & 0 \\ \dots & \dots & \dots & \dots & \dots \\ \frac{f^{(n-1)}(\lambda)}{(n-1)!}, & \dots & \dots & \dots & f(\lambda) \end{vmatrix}$$

Since $|\lambda| < \rho$, $f(z)$ and all its derivatives converge for $z = \lambda$. Hence $f(X)$ converges. This proves (5.6) and therefore also (5.5).

(5.8) Returning now to the arbitrary matrix X we note that we can define in particular

$$e^X = E + \frac{X}{1!} + \frac{X^2}{2!} \dots$$

for every X, and also

$$\log (E + X) = \frac{X}{1} - \frac{X^2}{2} + \dots$$

when the characteristic roots are less than one in absolute value.

(5.9) The usual rules for adding, multiplying, differentiating and generally combining series in X hold here also. However those for multiplying series in X by series in Y hold only when X and Y are commutative. Thus we may prove $e^{X+Y} = e^X \cdot e^Y$ when X and Y commute, but not so in the contrary case.

(5.10) *If* $f(X)$ *converges and* $Y = PXP^{-1}$, $|P| \neq 0$, *then* $f(Y)$ *converges also* (4.4) *and* $f(Y) = P(f(X))P^{-1}$.

6. (6.1) *If* $|X| \neq 0$ *there is a* Y *such that* $e^Y = X$.

Since Y need not be unique we do not insist on designating it by $\log X$.

Referring to (5.4) if $X = \text{diag}(X_1, X_2)$ and if we can find Y_1, Y_2 such that $X_1 = e^{Y_1}$ then $X = \text{diag}(e^{Y_1}, e^{Y_2})$ and so $Y = \operatorname{diag}(Y_1, Y_2)$ answers the question. Hence as in the proof of (5.5) we need only consider X of the type (3.6.1). Here $|X| \neq 0 \longrightarrow \lambda \neq 0$. In the same notations as before $X = \lambda E + Z = \lambda(E + \frac{1}{\lambda} Z)$. Since Z is nilpotent so is $\frac{1}{\lambda} Z$, and therefore we may define by (5.5) the function

$$Y_1 = \log (E + \frac{1}{\lambda} Z).$$

By (5.9):

$$E + \frac{1}{\lambda} Z = e^{Y_1}, \qquad X = \lambda E + Z = \lambda e^{Y_1}.$$

Since $\lambda \neq 0$ we may find a scalar μ such that $\lambda = e^{\mu}$. Then

$$X = (Ee^{\mu})\, e^{Y_1} = (e^{\mu E}) \cdot e^{Y_1} = e^{\mu E + Y_1}.$$

Therefore $Y = \mu E + Y_1$ answers the question.

(6.2) <u>If the series</u> (5.3) <u>for</u> f(X) <u>converges then the determinant</u>

(6.2.1) $$|f(X)| = \prod f(\lambda_j).$$

<u>Hence</u>

(6.2.2) $$|e^X| = e^{\sum \lambda_j} = e^{\operatorname{tr} X}.$$

<u>Thus</u> $Y = e^X$ <u>is never singular, and so it has an inverse</u> Y^{-1}. <u>Now</u> e^{-X} <u>exists likewise and</u> $e^X \cdot e^{-X} = e^{-X}e^X = E$, <u>so</u> $Y^{-1} = e^{-X}$.

Referring to (3.6) X is similar to a matrix

$$Y = \left\| \begin{matrix} \lambda_1 & & & \\ \alpha_{12} & \lambda_2 & & \\ & \ddots & \ddots & \\ & & \alpha_{n-1,n}, & \lambda_n \end{matrix} \right\|$$

with terms above the diagonal all zero. The λ_j are the characteristic roots each repeated as often as its multiplicity. This follows immediately from $|Y - \lambda E| = \prod (\lambda_j - \lambda)$.

By (4.4) $f(X) \sim f(Y)$ and so we may replace everywhere X, by Y, i.e. we merely need to prove (6.2) for Y. Now f(Y) is of the same form as Y with λ_j replaced by $f(\lambda_j)$. This implies that the $f(\lambda_j)$ are the characteristic roots of f(Y). Since the determinant is the product of the characteristic roots (6.2.1) holds for Y, hence also for X, and the rest follows.

7. (7.1) <u>Matrix functions of scalars</u>. Let

$$X = \| x_{ij}(t) \|$$

be an $m \times n$ matrix whose terms are [real or complex] scalar functions of a [real or complex] variable t differentiable over a certain range R. Under the rules of operation on matrices we have, if Δ denotes increments:

$$\frac{\Delta X}{\Delta t} = \left| \frac{\Delta x_{ij}}{\Delta t} \right| .$$

Hence if $\lim \frac{\Delta X}{\Delta t}$ exists it is defined as the derivative of X, written $\frac{dX}{dt}$, and exists over R, its expression being

$$\frac{dX}{dt} = \left\| \frac{dx_{ij}}{dt} \right\| .$$

The rules for the derivatives of scalars, for addition and for multiplication by scalars follow as usual. Similar limit arguments yield the definition of the Riemann integral:

$$Y(t) = \int_{t^0}^{t} X \, dt = \left\| \int_{t^0}^{t} x_{ij} \, dt \right\| .$$

If the x_{ij} are continuous on the path of integration then evidently

$$\frac{dY}{dt} = X.$$

(7.2) Suppose now that $X(t)$, $Y(t)$ are any two square matrices of the same order, differentiable over the same range R. Then XY is differentiable over the same range and an elementary argument yields

$$\frac{d(XY)}{dt} = X\frac{dY}{dt} + \frac{dX}{dt}Y. \tag{7.3}$$

Care must be taken here to keep X, Y always in the same order. From (7.3) we deduce readily

$$\frac{d(X_1 \ldots X_r)}{dt} = \sum X_1 \ldots X_{q-1}\frac{dX_q}{dt}X_{q+1} \ldots X_r \tag{7.3.1}$$

and therefore

$$\frac{dX^r}{dt} = \sum X^{q-1}\frac{dX}{dt}X^{r-q}. \tag{7.3.2}$$

If we differentiate both sides of $XX^{-1} = E$, $X \neq 0$, we obtain

$$\frac{d(X^{-1})}{dt} = -X^{-1}\frac{dX}{dt}X^{-1}.$$

(7.4) Observe explicitly that the application to $1 \times m$ matrices yields the derivatives and integrals of vector functions of scalars.

(7.5) If all the $x_{ij}(t)$ are continuous or analytic at a point or a given set, we will say for convenience that $X(t)$ is continuous or analytic at the same point or on the same set.

(7.6) Let A be a constant square matrix and set

(7.6.1) $X(t) = e^{tA} = e^{At} = E + \frac{t}{1!} A + \frac{t^2}{2!} A^2 + \dots .$

By differentiation we obtain

(7.6.2) $$\frac{dX}{dt} = Ae^{At}.$$

Notice that owing to the form of (7.3.2) we can only prove $\frac{d}{dt} e^{X(t)} = e^{X(t)} \frac{dX}{dt}$, when X and $\frac{dX}{dt}$ commute.

Combining (7.6.2) with (7.3), and setting for convenience $\frac{d}{dt} = D$, we obtain for any matrix X the analogue of the well known elementary relation:

(7.7) $$(D - A)X = e^{At} \cdot D \cdot e^{-At} X.$$

As an application consider the matrix differential equation

(7.8) $$\frac{dX}{dt} = AX, \text{ A constant.}$$

Owing to (7.7) it reduces to

$$e^{At} \cdot D \cdot e^{-At} X = 0.$$

Multiplying both sides by e^{-At} (see 6.2) we have

$$De^{-At} \cdot X = 0$$

and hence $e^{-At} \cdot X = C$, an arbitrary constant matrix. Hence the complete solution of (7.8) is $X = e^{At} \cdot C$. We will return to this later.

§2. VECTOR SPACES

8. We will assume familiarity with the first concepts: dimension, base, coordinates relative to a base. When the scalars are all real [complex] the space is said to be a real [complex] vector space. The only vector spaces which we shall encounter are finite dimensional. Let $\mathcal{V}$ be such a space. Its vectors will be denoted by an arrow over small latin characters with possible super-

scripts as: $\vec{a}$, $\vec{x}$, $\vec{a}^i$, ... Let in particular $\{\vec{e}^1, \ldots, \vec{e}^n\}$ be a base for $\mathcal{V}$. If we denote by $\vec{x}$ any element of $\mathcal{V}$ then we will have

$$(8.1) \qquad \vec{x} = \overrightarrow{x_1 e}^1 + \ldots + \overrightarrow{x_n e}^n.$$

The x_i are the coordinates or components of $\vec{x}$. If we adopt $\vec{x}$, x_i for the arbitrary vectors and their coordinates we will often denote $\mathcal{V}$ by $\mathcal{V}_x$. Similarly if say vectors and coordinates were $\vec{u}$, u_i we would write $\mathcal{V}_u$ for the space. In $\mathcal{V}_x$ the coordinates of $\vec{x}^h$ will be usually written x_{ih} (exceptionally and then explicitly stated x_{hi}).

The metrization of $\mathcal{V}_x$ will be done in the customary manner by means of a norm $\| \vec{x} \|$. We choose here for convenience

$$(8.2) \qquad \| \vec{x} \| = \sum |x_i|$$

and accordingly define the distance in $\mathcal{V}_x$ as

$$(8.3) \qquad d(x,x') = \| (\vec{x} - \vec{x}')\| = \sum |x_i - x_i'| \ .$$

As is well known this distance has the usual properties:

$$d(\vec{x},\vec{x}') = 0 \longrightarrow \vec{x} = \vec{x}';$$
$$d(\vec{x},\vec{x}') = d(\vec{x}',\vec{x}) \geq 0;$$
$$d(\vec{x},\vec{x}'') \leq d(\vec{x},\vec{x}') + d(\vec{x}',\vec{x}'').$$

With this specification of distance $\mathcal{V}_x$ is turned into a complete metric space which is topologically Euclidean space. We may show in fact that the above distance-function induces the same topology as the Euclidean distance $[\sum (x_i - x_i')^2]^{1/2}$. The completeness property of $\mathcal{V}_x$ implies that every Cauchy sequence has a limit.

(8.4) Let $\{\vec{e}^i\}$ be a base for $\mathcal{V}_x$. A square matrix A of order n defines a linear transformation of $\mathcal{V}_x$ into itself whereby $\vec{x}$ goes into $\vec{x}'$ designated by $A\vec{x}$, and whose coordinates are given by

$$(8.4.1) \qquad x_i' = \sum a_{ij} x_j.$$

If we identify $\vec{x}$ with the one-column matrix of its coordinates then $A\vec{x}$ is merely matrix multiplication.

Consider now a new n-space $\mathcal{V}_y$ referred to a base $\{\vec{f}^i\}$ and let P be a linear transformation $\mathcal{V}_y \rightarrow \mathcal{V}_x$ whereby $\vec{y}$ goes into $\vec{x}$ whose coordinates are given by

$$x_i = \sum p_{ij} y_j$$

or in matrix notation $\vec{x} = P\vec{y}$. If $\vec{x}'$ goes into $\vec{y}$ then $\vec{x}' = P\vec{y}'$ and so, $\vec{x}$, $\vec{x}'$ being related as before we have $P\vec{y}' = AP\vec{y}$. Assuming now P non-singular we will have $\vec{y}' = P^{-1}AP\vec{y}$. In other words the transformation of $\mathcal{V}_x$ into itself by A corresponds to a transformation of $\mathcal{V}_y$ into itself by a matrix $\sim A$. Clearly also every matrix $\sim A$ is related to A in this manner. We may therefore interpret the properties of A invariant under similitude, as those of the transformations which are invariant under a non-singular linear transformation from space to space.

(8.5) Let $A \sim \text{diag}\,(A_1, \ldots, A_r)$ where the A_i are like (3.6.1), and let $\{\vec{f}^i\}$ be the base such that on passing to it A goes into the canonical form. Denote by ρ_i the order of A_i and set $\sigma_i = \rho_1 + \ldots' + \rho_i$. Let also $A_i^* = \text{diag}\,(0, \ldots, 0, A_i, 0, \ldots, 0)$. The units

$\vec{f}^{\sigma_{i-1}+1}, \ldots, \vec{f}^{\sigma_i}$ may be characterized as follows. First they are all annulled by the A_j^*, $j \neq i$. Then

$$(8.5.1) \qquad \begin{aligned} A_i^* \vec{f}^{\sigma_{i-1}+1} &= \lambda_i \vec{f}^{\sigma_{i-1}+1} \\ A_i^* \vec{f}^{\sigma_{i-1}+h} &= \vec{f}^{\sigma_{i-1}+h-1} + \lambda_i \vec{f}^{\sigma_{i-1}+h}, \quad h > 1, \end{aligned}$$

which provides a complete description.

Suppose in particular A real and among the λ_j let there be found r pairs which are complex conjugates which for convenience we may assume to be $(\lambda_1, \bar{\lambda}_1), \ldots, (\lambda_r, \bar{\lambda}_r)$ and the rest $\lambda_{2r+1}, \ldots, \lambda_n$ real. If we obtain the vectors $\vec{f}$ for $\lambda_1, \ldots, \lambda_r$ then their conjugates will do for $\bar{\lambda}_1, \ldots, \bar{\lambda}_r$. The coordinates of a real vector referred to this base will then be $x_1, \ldots, x_r, \bar{x}_1, \ldots, \bar{x}_y, x_{2r+1}, \ldots, x_n$ where the x_{2r+i} are real. We will then say that the coordinate system is <u>real</u>.

(8.6) If $\vec{x}(t)$ depends upon t then the derivatives and integrals of $\vec{x}(t)$ may be defined in the usual manner and are written

$$\frac{d\bar{x}}{dt}, \quad \int_{t^0}^{t} \vec{x}(t)dt.$$

Both are vectors, their components being respectively

$$\frac{dx_i}{dt}, \quad \int_{t_0}^{t} x_i(t)dt.$$

Clearly

$$\frac{d}{dt} \int_{t_0}^{t} \vec{x}(t)dt = \vec{x}(t).$$

We also note the following useful inequality. Suppose t real and $\| x(t) \| < M$ for $t_0 \leq t \leq t_1$. Then

$$(8.6.1) \qquad \left\| \int_{t_0}^{t_1} x(t)dt \right\| < n\, M\, |t_1 - t_0| .$$

(8.7) Let $\vec{y} = (y_1, \ldots, y_r)$ be a real or complex vector in some $\mathcal{V}_y$. Suppose the x_i analytic in the y_j at $\vec{y}^0 = (y_1^0, \ldots, y_r^0)$, i. e. representable by power series in the y_j valid in a neighborhood of $\vec{y}^0$ in $\mathcal{V}_y$. We will then say: $\vec{x}$ is analytic in $\vec{y}$ at $\vec{y}^0$. This is the prototype of a readily understood terminology used extensively later.

(8.8) In dealing with n dimensional spaces $\mathcal{V}_x$ it will be convenient to define as a sphere of center $\vec{x}^0$ and radius ρ, written $\mathcal{S}(x^0, \rho)$ the set $\| \dot{x} - x^0 \| < \rho$.

(8.9) Frequently besides the vector variables $\vec{x}$, $\vec{y}$, ..., there will occur a real parameter t referred to as the time, and whose range is the real line L. Instead of $\mathcal{V}_x$ for instance we shall have a product space $\mathcal{W} = L \times \mathcal{V}_x$ and new spheres $\sum(\vec{x}^0, t^0, \rho)$ defined by $\|\vec{x} - \vec{x}^0\| + |t - t^0| < \rho$.

The relation of the spheres to open sets, limits, etc., are as in real variables and need not be discussed here.

(8.10) Whenever $\mathcal{V}_x$ is two-dimensional it will be convenient to revert to the more usual "spheres", namely the circular regions of Euclidean geometry. As is well known this does not affect the standard concepts of open sets, ...

§3. ANALYTIC FUNCTIONS OF SEVERAL VARIABLES

9. We shall have repeated occasion to consider analytic functions of several real or complex variables as well as mixed functions analytical in some, but not in all, the variables.

(9.1) Consider first $\mathcal{V}_x$ complex. Write $\vec{x} = \vec{y} + i\vec{z}$, viz. $x_j = y_j + iz_j$ for $j = 1, \ldots, n$. A function $f(\vec{x})$ is said to be analytic in a region Ω of $\mathcal{V}_x$ if it has first partial derivatives relative to all y_j and z_j which are continuous in $\vec{y}$ and $\vec{z}$ at all points of Ω and if it satisfies the Cauchy-Riemann differential equations relative to each pair of y_j and z_j at all points of Ω. The function f is said to be <u>holomorphic</u> in Ω if it is analytic and single-valued in Ω; f is said to be analytic or holomorphic in a closed set F in $\mathcal{V}_x$ if it is analytic or holomorphic is some neighborhood of F, (some region $\supset$ F).

A n.a.s.c. for analyticity in Ω is that f may be expanded in Taylor series around each point $\vec{\xi}$ of Ω. The series will be convergent in a set

$$\mathcal{J}(\vec{\xi}, \alpha) : |x_j - \xi_j| < \alpha, \qquad j = 1, 2, \ldots, n,$$

which we call conveniently a toroid of center $\vec{\xi}$ and radius α. Moreover α may be chosen such that the toroid is in Ω.

(9.2) Suppose now $\mathcal{V}_x$ real and let $[\mathcal{V}_x]$ be its complex extension, i. e. the complex vector space obtained by allowing the coordinates x_j to take complex values. A real function $f(\vec{x})$ will be said to be analytic or holomorphic in a region Ω of $\mathcal{V}_x$ under the following conditions: there exists a region $[\Omega]$ of $[\mathcal{V}_x]$ and a function $[f]$ on $[\mathcal{V}_x]$ analytic or holomorphic in $[\Omega]$, and such that: (a) $[\Omega] \supset \Omega$; (b) the values of $[f]$ on Ω are those of f.

(9.3) The definitions just given for the real case may seem indirect. They have however the advantage to guarantee that the following important property holds:

(9.4) Theorem. *If a series of real or complex functions analytic or holomorphic in a region Ω, is uniformly convergent in Ω, then the limit is analytic or holomorphic in Ω.*

For the complex case this is a standard theorem due to Weierstrass (see Osgood II p. 15) and for the real case it is a consequence of our definition.

(9.5) An analytical vector is a vector $\vec{f}(\vec{x})$ whose components $f_i(x_1, \ldots, x_n)$ are analytical.

(9.6) Given two series

$$a = a_1 + a_2 + \cdots, \qquad b = b_1 + b_2 + \cdots,$$

the second is said to be a majorante of the first, written $a \ll b$ (Poincaré's notation) whenever $|a_m| \leq |b_m|$ for every m. More generally, if the multiple series

$$a = \sum a_{m_1,\ldots,m_p}, \qquad b = \sum b_{m_1,\ldots,m_p}$$

are such that $|a_{m_1,\ldots,m_p}| \leq |b_{m_1,\ldots,m_p}|$ for every combination $m_1,\ldots,m_p$ then b is called a majorante of a, written as before $a \ll b$.

(9.6.1) If $m = \sum m_i$ then it is often convenient to denote by (m) the set $\{m_1, \ldots, m_p\}$. Thus we would write above:

$$a = \sum a_{(m)}, \qquad b = \sum b_{(m)}. \tag{9.6.2}$$

(9.7) Suppose that $F(x_1, \ldots, x_n)$ is holomorphic in the closed region Ω : $|x_i| \leq A_i$, $i = 1, 2, \ldots, n$, where the A_i are positive constants. Since Ω is compact $|F|$ has an upper bound M in Ω. It is then shown in the treatises on the subject (see for instance Picard, Traité d'Analyse, vol. III, Ch. 9) that F admits in Ω the McLaurin expansion

$$F = \sum F^{(m)} x_1^{m_1} \ldots x_n^{m_n}$$

with the following estimate for the coefficients:

$$F^{(m)} < \frac{M}{A_1^{m_1} \ldots A_n^{m_n}}. \tag{9.8}$$

If we identify F with the series we have therefore

$$F \ll \frac{M}{\prod (1 - \frac{x_i}{A_i})}. \tag{9.9}$$

If $A = \inf A_i$ then another useful relation of the same type is

$$F \ll \frac{M}{1 - \frac{1}{A} \sum x_i}. \tag{9.10}$$

It is in fact a simple matter to show that

$$\frac{1}{\prod (1 - \frac{x_i}{A_i})} \ll \frac{1}{1 - \frac{1}{A} \sum x_i}$$

from which (9.10) follows

10. Consider a transformation ϕ from the real space $\mathcal{V}_x$ to the real space $\mathcal{V}_y$ represented by

$$y_i = \phi_i(x_1, \ldots, x_n). \tag{10.1}$$

We will say that ϕ is regular at $\vec{x}^0$ whenever the ϕ_i are analytic at the point and the Jacobian

$$J = \left| \frac{\partial \phi_i}{\partial x_j} \right|$$

does not vanish at the point. The implicit function theorem for the system (10.1) yields immediately the following property:

(10.2) *Let ϕ be regular at $\vec{x}^0$ and let $\vec{y}^0 = \phi\vec{x}^0$. There can be chosen in $\mathcal{V}_x$, $\mathcal{V}_y$ n-cells E^n_x, E^n_y containing respectively $\vec{x}^0$, $\vec{y}^0$ and such that ϕ maps topologically E^n_x onto E^n_y.*

(10.3) Certain analytical loci will occur on repeated occasions in the sequel. As the most general loci of this type will not be needed we shall confine our remarks to certain simple types.

Consider a system of relations with $p \leq n$:

$$x_i - x_i^0 = \phi_i(u_1, \ldots, u_p) \quad (i = 1,2,\ldots,n) \tag{10.3.1}$$

where the ϕ_i are real, vanish at $\vec{u} = 0$, are holomorphic in a certain p-cell $E^p_u : \| \vec{u} \| < \rho$, and have a Jacobian matrix of rank p at $\vec{u} = 0$. We consider (10.3.1) as defining a mapping ϕ of E^p_u into $\mathcal{V}_x$. It is a consequence of our assumptions that ϕ is a topological mapping of E^p_u into a p-cell E^p through $\vec{x}^0$ in $\mathcal{V}_x$. The cell E^p is known as *an elementary analytical* p-*cell* through $\vec{x}^0$. When p = 1 we also speak of an elementary *analytical arc* through the point. An *analytical curve* Λ in $\mathcal{V}_x$ is a connected set such that every point has a neighborhood which is an elementary arc. That is to say every point $\vec{x}^0$ of Λ has a neighborhood E^1 which admits a parametric representation

(10.3.2) $$x_i - x_i^0 = \psi_i(v),\ \psi_i(0) = 0, \qquad |v| < \alpha,$$

where the ψ_i are analytical on their range and the derivatives $\psi_i'(v)$ never vanish simultaneously on the range. When the analytical curve λ is compact it is a Jordan curve, otherwise it is an arc.

The following intersection property will be found useful later.

(10.4) Let E_u^p and the analytical curve λ have an infinite number of intersections with a limit-point P contained in both. Then P has a neighborhood ν in the intersection μ which is a subarc of λ.

We may suppose the representations as above with P as $\vec{x}^0$. Under the assumption we may solve (10.3.1) for $u_1, \ldots, u_p$ in terms of p of the differences $x_i - x_i^0$, say in terms of the first p differences and thus replace the system in the vicinity of P by another of the form

$$x_{p+i} - x_{p+i}^0 = f_i(x_1-x_1^0, \ldots, x_p-x_p^0)$$

where the $f_i(w_1, \ldots, w_p)$ are analytic at the origin $\vec{w} = 0$. As a consequence the analytic functions of v

$$F_i(v) = f_i(\psi_1(v), \ldots, \psi_p(v)) - \psi_{p+i}(v)$$

have an infinity of zeros in the vicinity of $v = 0$. It follows that F_i vanishes on some interval $|v| < \alpha_i$. If α is the least α_i then P has for neighborhood in μ the subarc of λ corresponding to $|v| < \alpha$.

(10.5) Application. Two analytical curves λ, μ which have an infinite number of intersections with a limit-point P in both λ and μ, necessarily coincide.

Let ξ be the intersection. The point P is in an elementary subcell E_u^1 of λ, and μ intersects E_u^1 in an infinite number of points. Hence P is contained in a subarc ν of μ which is in the intersection ξ and is a neighborhood of P in both λ and μ. Let η be the largest subarc of both λ and μ which contains ν and suppose that ν has an end point Q. Since ξ is closed in both λ and μ, Q is in ξ. By the above argument Q is contained in some subarc ν' of both λ and μ. Since this contradicts the maximal property of η, Q cannot exist. Hence $\eta = \lambda = \mu$. This proves (10.5).

(10.6) By an <u>analytical</u> (n-1)-<u>manifold</u> M^{n-1} in $\mathcal{V}_x$ we shall understand a locus represented by an equation

$$(10.6.1) \qquad f(x_1, \ldots, x_n) = 0$$

where f is real analytical in a certain region Ω of $\mathcal{V}_x$ and the partials $\frac{\partial f}{\partial x_i}$ do not vanish simultaneously in Ω. Under this assumption if $\vec{x}^0$ is a point of M^{n-1} and say $\frac{\partial f}{\partial x_n^0} \neq 0$, one may solve (10.6.1) in the neighborhood of $\vec{x}^0$ for $x_n - x_n^0$ as

$$x_n - x_n^0 = \phi(x_1, \ldots, x_{n-1})$$

where ϕ is analytical in the vicinity of $(x_1^0, \ldots, x_{n-1}^0)$. Thus the $x_i - x_i^0$ may be expressed in the form (10.3.1) with $p = n-1$ and $u_h = x_h - x_h^0$, $h = 1, 2, \ldots, n-1$. In other words $\vec{x}^0$, i.e. any point of M^{n-1} has a neighborhood in M^{n-1} which is an elementary analytical (n-1)-cell.

When n = 2, M^1 is merely an analytical curve in the plane $\mathcal{V}_x$.

11. <u>On a question of reality</u>. Largely as a consequence of the possible complex characteristic roots of real matrices we shall find it convenient to describe certain systems as <u>real</u> in a broader sense than commonly understood.

Consider an n-dimensional vector space whose elements are sets $x_1, \bar{x}_1, \ldots, x_r, \bar{x}_r, x_{2r+1}, \ldots, x_n, x_{2r+j}$ is real. The resulting vector space $\mathcal{V}_x$ is still referred to as a <u>real n dimensional vector space</u>.

A function $f(x_1, \ldots, x_n) = f(\vec{x})$ is said to be real whenever $f(x_1, \overline{x}_1, \ldots, x_n) = f(\overline{x}_1, x_1, \overline{x}_2, x_2, \ldots, x_n)$. It is said to be analytic at a point $\vec{x}^0$ whenever f is analytic in $x_1, \overline{x}_1, \ldots, x_n$, (all considered as independent variables) about $(x_1^0, \overline{x}_1^0, \ldots, x_n^0)$. A vector function $\vec{f}(\vec{x}) = (f_1(\vec{x}), \ldots, f_n(\vec{x}))$ is said to be real whenever the f_{r+i} are real and

$$f_i(\overline{x}_1, x_1, \ldots, x_n) = \overline{f}_i(x_1, \overline{x}_1, \ldots, x_n), \qquad i \leq r.$$

It is said to be analytic whenever the f_i are analytic.

To illustrate our general meaning, take $r = 1$, $n = 3$. Thus the coordinates may be named $x, \overline{x}, y$. The function $f(x, \overline{x}, y) = f(\overline{x}, x, y)$. It is analytic at $(x^0, \overline{x}^0, y^0)$ whenever f is analytic in $x, \overline{x}, y$ in the ordinary sense at the point. A real vector function $\vec{f}$ is a triple $(f, \overline{f}, g)$ of functions of $x, \overline{x}, y$ where g is real and $\overline{f}(x, \overline{x}, y) = f(\overline{x}, x, y)$.

12. Topological and related considerations. Free use will be made in the sequel of certain symbols and concepts of set theory and topology. A few words concerning these may help the reader unfamiliar with them.

(12.1) First as to set theoretic symbols. The following should be kept in mind. If $A_1, \ldots, A_r$ or $\{A_\alpha\}$ are sets or a collection of sets then

$A_1 \cup \ldots \cup A_r$ or $\cup A_\alpha$ denotes their union (all the elements in one of the sets);

$A_1 \cap \ldots \cap A_r$ or $\cap A_\alpha$ denotes their intersection (all the elements common to all the sets).

If A, B are two sets then $A \subset B$ or $B \supset A$ mean: A is a subset of B.

If A is the set of all elements $\{a_\alpha\}$ then $a_\alpha \in A$ or $A \ni a_\alpha$ means a_α is an element of A.

(12.2) Coming now to topological notions it is to be understood throughout that all sets under consideration are Euclidean (subsets of Euclidean spaces).

It will be recalled that a mapping f is a continuous and single-valued transformation or function. A topological transformation or mapping is one whose inverse is also a mapping: f is one-one and bicontinuous. A topological image of a set is its image under a topological mapping. Notable examples are:

The open [closed] arc is the topological image of the interval $0 < u < 1$ [of the segment $0 \leq u \leq 1$]. The term "arc" will generally refer to the open arc.

The n-cell is the topological image of the so-called "n-dimensional interval", namely the subset of Euclidean n-space (with coordinates u_i) defined by

$$0 < u_i < 1, \qquad i = 1, 2, \ldots, n.$$

The closed n-cell is the topological image of

$$0 \leq u_i \leq 1, \qquad i = 1, 2, \ldots, n.$$

In a metric space $\mathcal{R}$ i. e. a space with specified distance $d(x,y)$ we shall understand by sphere of center x_0 and radius ρ, written $\mathcal{S}(x_0,\rho)$ (or sometimes $\Sigma(x_0,\rho)$) the set of all points y of $\mathcal{R}$ such that $d(x_0,y) < \rho$. In an Euclidean n-space the spheres are all n-cells.

In a metric space $\mathcal{R}$ an open set U is a set such that if $x_0 \in U$ then some $\mathcal{S}(x_0,\rho) \subset U$. An arbitrary union of open sets is an open set. The complement $F = \mathcal{R} - U$ of an open set is a closed set in $\mathcal{R}$.

A neighborhood of a point x or of a set A in $\mathcal{R}$ is an open set U containing x or A.

The closure $\bar{A}$ of any set A in $\mathcal{R}$ is the intersection of all the closed sets containing A. It may also be defined as the set of all points whose neighborhoods all meet A.

A subset A of an Euclidean space $\mathcal{E}^n$ is itself metric, i. e. it may be assigned a distance $d(x,y)$ for any $x, y \in A$, namely their distance in $\mathcal{E}^n$. If A, B are two such metric sets with $d(x,y)$, $d'(x',y')$ as their distances then by the metric product or merely product $A \times B$ is meant the set of all couples (x,x') $x \in A$, $x' \in B$, with the distance defined by

$$D((x,x'), (y,y')) = d(x,y) + d'(x',y') .$$

Usually one of the two factors say B will be a real line $-\infty < u < +\infty$ or an interval or segment of that line, and the other A will be a portion of Euclidean space. The distance will then be

$$D((x,u), (x',u')) = d(x,x') + |u-u'|.$$

(12.3) Connectedness. A set A is connected whenever we cannot write $A = B \cup C$ where B and C are disjoint and both open and closed in A (intersections of closed or open sets with A). A non connected set A is a union of maximal connected subsets known as the components of A.

(12.4) Compact sets. Compactness is a topological property which may be defined as in real variables. It is sufficient as regards the sequel to recollect the following properties:

(12.5) A closed and bounded Euclidean set is compact. Thus a closed n-cell is compact.

(12.6) If a mapping of a compact set into a metric set is one-one and continuous then it is topological.

(12.7) Remark. For further details regarding the topological concepts scarcely touched upon here, see the author's Colloquium Lectures: Algebraic Topology, Ch. I.

CHAPTER II

DIFFERENTIAL EQUATIONS

§1. GENERALITIES

1. Let t be a real variable and $x_i(t)$ a finite set of functions of t over a certain range R, where we suppose that the requisite derivatives mentioned below all exist. By a <u>system of differential equations</u> or <u>differential system</u> in the unknowns $x_i(t)$ is meant a system of equations

$$(1.1)\qquad F_i\left(x_1, \ldots, x_n, \ldots, \frac{d^h x_i}{dt^h}, \ldots, t\right) = 0$$

in the x_i, their derivatives, and t.

If we introduce new unknowns as indicated by the rule

$$(1.2)\qquad \frac{dx_i}{dt} = y_{i1}, \quad \frac{d^2x_i}{dt^2} = \frac{dy_{i1}}{dt} = y_{i2}, \ldots,$$

we may replace (1.1) by an equivalent system (generally with more equations) of the form

$$(1.3)\qquad G_i\left(y_1, \ldots, y_m, \frac{dy_i}{dt}, t\right) = 0,$$

in which G_i contains just one derivative.

(1.4) <u>Example</u>. The single differential equation

$$(1.4.1.)\qquad \frac{d^n x}{dt^n} + a_1(x,t)\frac{d^{n-1}x}{dt^{n-1}} + \ldots + a_n(x,t) = 0$$

may be replaced by

$$x = x_1, \frac{dx_1}{dt} = x_2, \ldots, \frac{dx_{n-1}}{dt} = x_n, \tag{1.4.2}$$
$$\frac{dx_n}{dt} + a_1(x_1,t)x_{n-1} + \ldots + a_n(x_1,t) = 0.$$

2. Systems of the generality of (1.1) have been investigated by various authors, the latest and most thorough being J. F. Ritt. Our program, far more modest, will be confined to systems in which the derivatives occur linearly, that is to say, to systems of one or more equations (1.4.1), or of the form (1.4.2). In fact, written more explicitly, we shall deal primarily with systems of the form

$$\frac{dx_i}{dt} = p_i(x_1, \ldots, x_n,t), \qquad (i = 1,2, \ldots, n) \tag{2.1}$$

admitting generally all real values (more exceptionally all complex values) as the range of functions and variables other than t. It is for such systems alone that sufficiently extensive existence theorems are available.

The independent variable t will often be referred to as the <u>time</u>. This is justified on the ground that many systems of differential equations arise from problems in dynamics or other branches of mathematical physics.

3. (3.1) Returning to the basic system (2.1) let $(x_1, \ldots, x_n)$ be considered as a vector $\vec{x}$ in a space $\mathcal{V}_x$. Then $(p_1, \ldots, p_n)$ defines a vector-function of $\vec{x}$ and t, conveniently written $\vec{p}(\vec{x},t)$. Thus (2.1) assumes the simple form

$$\frac{d\vec{x}}{dt} = \vec{p}(\vec{x},t). \tag{3.2}$$

The vectors $\vec{p}(\vec{x},t)$ form a field F(t) in $\mathcal{V}_x$ called the <u>field</u> of (3.2) and (3.2) may be interpreted as requiring to find a vector $\vec{x}$ whose derivative $\frac{d\vec{x}}{dt}$ defines a certain preassigned field F(t). Of course if $\vec{p}(\vec{x},t)$ is a function $\vec{p}(\vec{x})$ then the field F is constant.

Since $\vec{p}(\vec{x},t) \in \mathcal{U}_x$, the metric of $\mathcal{U}_x$ may be applied to $\vec{p}$, and $\| \vec{p}(\vec{x},t) \|$, $\| \vec{p}(\vec{x},t)-\vec{p}(\vec{x}',t)\|$ have a definite meaning. To say that "$\vec{p}$ is bounded under certain conditions," or that "$\vec{p}$ has an upper bound M" is to be interpreted as "$\| \vec{p} \|$ is bounded under certain conditions," "$\| \vec{p} \|$ has an upper bound M".

§2. THE FUNDAMENTAL EXISTENCE THEOREM.

4. In order to facilitate the statement it will be convenient to introduce a certain preliminary concept. We shall be considering a region Ω of the space $\mathcal{W}$ of $(\vec{x}, t)$ and a point $(\vec{\xi}, \eta)$ of that space. We shall understand by a box $B(\alpha,\tau)$ of center $(\vec{\xi}, \eta)$ an open set of $\mathcal{W}$ which is a product $U_\alpha \times I_\tau$ where U_α is the set $\| \vec{x}-\vec{\xi} \| < \alpha$ and I_τ is the time interval $|t-\eta| < \tau$. We may now state the

(4.1) Existence Theorem of Cauchy-Lipschitz for real functions. Let Ω be a region of the space $\mathcal{W}$ of $(\vec{x},t)$ such that:

(4.1.1) $\vec{p}(\vec{x},t)$ is continuous in Ω.

(4.1.2) for every pair of points $(\vec{x},t),(\vec{x}',t)$ in Ω there is fulfilled a Lipschitz condition

$$(4.1.3) \qquad \| \vec{p}(\vec{x},t) - \vec{p}(\vec{x}',t) \| < k \, \| \vec{x}-\vec{x}' \|, \qquad k > 0.$$

Corresponding to any $(\vec{\xi},\eta)$ in Ω and the various boxes being of center $(\vec{\xi},\eta)$, there exists a $B(2\alpha,2\tau) \subset \Omega$, where τ depends upon α, and with the following properties: For every $(\vec{x}^0,t^0) \in B(\alpha,\tau)$ there exists a unique solution $\vec{x}(t)$ of (3.2) defined over I_τ such that $(\vec{x}(t),t)$ remains in $B(2\alpha,\tau)$ when t is in I_τ, and that $\vec{x}(t^0) = \vec{x}^0$.

The proof will rest upon Picard's classical process of successive approximations.

We begin with a preliminary remark. We know that $(\vec{\xi},\eta)$ has a neighborhood in Ω of the form $B(2\alpha,2\tau)$. Now let M be an upper bound for $\vec{p}(\vec{x},t)$ in Ω. It is clear that

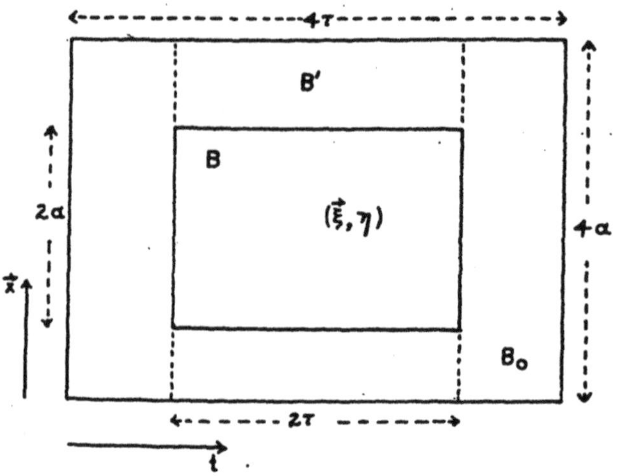

Fig. 1

$$\phi(\tau) = \frac{M}{k}\,(e^{k\tau}-1) > 0.$$

Since we may replace τ by a smaller number, we may take it such that $\phi(2\tau) < \alpha$, that is to say take $\tau < \frac{1}{2k}\log\,(1+\frac{\alpha k}{M})$.

We now pass to the proof proper. We shall take $(\vec{x}^0,t^0)$ as an arbitrary point in $B(\alpha,\tau)$. Thus $\|\vec{x}^0-\vec{\xi}\| < \alpha$, $|t^0-\eta| < \tau$.

(a) __Existence of__ $\vec{x}(t)$. Consider the sequence of vectors $\vec{x}^m(t)$ defined in succession by

$$(4.2)_m \qquad \vec{x}^{m+1} = \vec{x}^0 + \int_{t_0}^{t} \vec{p}(\vec{x}^m,t)dt, \qquad t \in I_\tau.$$

We will show that for $t \in I_\tau$, the sequence has a limit which satisfies (3.2).

For $t \in I_\tau$, we have first of all

$$|t-t^0| \leq |t-\eta| + |\eta-t^0| < \tau + \tau = 2\tau.$$

Then

$$\| \vec{x}^0 - \vec{x}^1 \| \leq M \, |t - t^0| < m \, . \, 2\tau < \phi(2\tau) < \alpha.$$

Hence

$$\| \vec{x}^1 - \vec{\xi} \| \leq \| \vec{x}^1 - \vec{x}^0 \| + \| \vec{x}^0 - \vec{\xi} \| < \alpha + \alpha = 2\alpha.$$

Therefore for $t \in I_\tau$, $(\vec{x}^1(t), t)$ remains in $B(2\alpha, \tau)$.

Now

$$\vec{x}^2 - \vec{x}^1 = \int_{t_0}^{t} (\vec{p}(\vec{x}^1, t) - \vec{p}(\vec{x}^0, t))dt.$$

By (4.1.2)

$$\| \vec{x}^1 - \vec{x}^2 \| \leq \left| \int_{t_0}^{t} \| \vec{p}(\vec{x}^0, t) - \vec{p}(\vec{x}^1, t \| \, dt \right|$$

$$\leq k \left| \int_{t_0}^{t} \| \vec{x}^0 - \vec{x}^1 \| dt \right| \leq kM \frac{|t - t_0|^2}{2} < kM \frac{(2\tau)^2}{2} .$$

Hence

$$\| \vec{x}^0 - \vec{x}^2 \| \leq \| \vec{x}^0 - \vec{x}^1 \| + \| \vec{x}^1 - \vec{x}^2 \| < m \, . \, 2\tau + kM \frac{(2\tau)^2}{2} <$$

$$\phi(2\tau) < \alpha.$$

Then

$$\| \vec{x}^2 - \vec{\xi} \| \leq \| \vec{x}^2 - \vec{x}^0 \| + \| \vec{x}^0 - \vec{\xi} \| < \alpha + \alpha = 2\alpha .$$

Therefore $t \in I_\tau \longrightarrow (\vec{x}^2(t), t) \in B(2\alpha, \tau)$.

Proceeding in the same manner we obtain

$$(4.3) \qquad \| \vec{x}^m - \vec{x}^{m+1} \| < Mk^m \frac{|t - t_0|^{m+1}}{(m+1)!} < \frac{M}{k} \frac{(2k\tau)^{m+1}}{(m+1)!}$$

Hence by repeated application of the triangle relation

$$\| \vec{x}^0 - \vec{x}^{m+1} \| < M\cdot 2\tau + Mk \frac{(2\tau)^2}{2!} + \ldots + Mk^m \frac{(2\tau)^{m+1}}{(m+1)!}$$

$$< \phi(2\tau) < \alpha$$

and so again $(\vec{x}^{m+1}(t),t) \in B(2\alpha,\tau)$ for $t \in I_\tau$.

It is clear from (4.3) that $\{\vec{x}^m(t)\}$, $t \in I_\tau$, is a Cauchy sequence. Since ϑ is complete, the sequence converges to a limit $\vec{x}(t)$ such that $\| \vec{x}(t)-\vec{x}^0 \| \leq \phi(2\tau) < \alpha$, $\| \vec{x}(t)-\vec{\xi} \| < 2\alpha$, and hence $\vec{x}(t) \in U_{2\alpha}$ and the convergency is uniform for $t \in I_\tau$. It follows then immediately that $\int_{t_0}^t \vec{p}(\vec{x}^m,t)dt$ converges uniformly to $\int_{t_0}^t \vec{p}(\vec{x},t)dt$ for $t \in I_\tau$. Now for $t \in I_\tau$ and all m, we have

$$\vec{x} - \vec{x}^0 - \int_{t_0}^t \vec{p}(\vec{x},t)dt$$

$$= \vec{x} - (\vec{x}^{m+1} - \int_{t_0}^t \vec{p}(\vec{x}^m,t)dt) - \int_{t_0}^t \vec{p}(\vec{x},t)dt$$

$$= \vec{x} - \vec{x}^{m+1} + \int_{t_0}^t \vec{p}(\vec{x}^m,t)dt - \int_{t_0}^t \vec{p}(\vec{x},t)dt$$

and therefore

$$\| \vec{x}-\vec{x}^0 - \int_{t_0}^t \vec{p}(\vec{x},t)dt \|$$

$$\leq \| \vec{x}-\vec{x}^{m+1} \| + \| \int_{t_0}^t \vec{p}(\vec{x}^m,t)dt - \int_{t_0}^t \vec{p}(\vec{x},t)dt \| .$$

Hence for $t \in I_\tau$,

$$\| \vec{x}-\vec{x}^0 - \int_{t_0}^t \vec{p}(\vec{x},t)dt \|$$

can be arbitrarily small (this being uniformly in t) for m sufficiently large, and therefore it is identically zero. Now we have, for $t \in I_\tau$,

$$(4.4)\qquad \vec{x} = \vec{x}^0 + \int_{t_0}^{t} \vec{p}(\vec{x},t)\, dt,$$

which shows that $\vec{x} = \vec{x}(t)$ satisfies (3.2) and $\vec{x}(t^0) = \vec{x}^0$.

(b) Uniqueness of $\vec{x}(t)$. Suppose that $\vec{y}(t)$ defines a second solution over $I_{\tau'}$: $|t-\eta| < \tau'$ such that $\vec{y}(t^0) = \vec{x}^0$. We may suppose $\tau \leq \tau'$, and so the second solution is defined likewise over I_τ. Since $\vec{x}(t)$, $\vec{y}(t)$ are differentiable they are continuous. Hence given any $\beta > 0$ there will exist a positive $\sigma < \tau$ such that $|t-t^0| < \sigma \longrightarrow \| \vec{x}-\vec{x}^0 \|$, $\| \vec{y}-\vec{x}^0 \| < \beta$ and therefore $\| \vec{x}-\vec{y} \| < 2\beta$. Now by (4.1.2):

$$(4.5)\qquad \| \vec{x}-\vec{y} \| \leq k \left| \int_{t_0}^{t} \| \vec{x}-\vec{y} \|\, dt \right| < 2\beta k\, |t-t^0| .$$

By repeated substitution in (4.5) we obtain

$$\| \vec{x}-\vec{y} \| < 2\beta \frac{(k|t-t^0|)^m}{m!} \longrightarrow 0 \text{ with } \frac{1}{m} .$$

Hence $\| \vec{x}-\vec{y} \| = 0$ and therefore $\vec{x} = \vec{y}$ for $|t-t^0| < \sigma$. Thus if $\vec{x}(t)$, $\vec{y}(t)$ assume the same value for any $t^0 \in I_\tau$, they coincide on an interval containing t^0 and contained in I_τ.

Now consider the maximal interval $I' : t' < t < t''$ containing t^0, contained in I_τ and such that $\vec{z}(t) = \vec{x}(t) - \vec{y}(t) = 0$ on I'. If $I' = I_\tau$, then $\vec{x}(t) = \vec{y}(t)$ on I_τ, and the proof is complete. In the contrary case, one of t' or t'', say t' is in I_τ. Since $\vec{z}(t)$ is continuous in I_τ and is zero in I', it is also zero at t', since $t' \in \bar{I}'$. It follows from the earlier argument that $\vec{z}(t)$ vanishes in an interval I'' containing t'' and contained in I_τ. Thus we can augment I' by I'', and so we get a contradiction.

This completes the proof of the existence theorem.

5. Complementary Remarks. (5.1) Of the two basic conditions (4.1.1), (4.1.2) in the statement of the existence theorem the second is less natural than the first, and would seem more difficult to verify. The following

property will therefore be useful in this direction, and also will suffice for later applications.

(5.2) <u>If the partial derivatives</u> $\partial p_i/\partial x_j$ <u>exist and are continuous and bounded in</u> Ω <u>then</u> (4.1.3) <u>holds</u>.

For we may clearly replace Ω by a box $B(\beta,\tau)$. Applying now the mean value theorem to p_i, with $x, x' \in U_\beta$ we find

$$p_i(\vec{x},t) - p_i(\vec{x}',t) = \sum \frac{\partial p_i}{\partial x_j}(\vec{x}^*,t)(x_j-x_j'),$$

where $\vec{x}^* \in U_\beta$. Hence if A is an upper bound of the $\left|\frac{\partial p_i}{\partial x_j}\right|$ for all p_i, we find

$$\| \vec{p}(\vec{x},t) - \vec{p}(\vec{x}',t) \| < nA \,.\, \| \vec{x}-\vec{x}' \| \,.$$

Thus (4.1.2) holds with $k = nA$.

(5.3) It is worth while to exhibit functions which do not satisfy a Lipschitz condition, e.g. (a) $\vec{p}(\vec{x},t)$ with $p_i = \frac{1}{x_i}$ for $0 < x_i < 1$; (b) $\vec{p}(\vec{x},t)$ with $p_i = \sqrt{x_i}$ for $0 < x_i$.

(5.4) <u>Domain of continuity</u>. The union of all the sets such as Ω is an open set of $\mathcal{W}$. We choose one of its components $\mathcal{D}$ and call it the <u>domain of continuity of the differential equation</u> (3.2). Henceforth the equation is supposed to be taken together with a definite domain $\mathcal{D}$ and all our operations will be generally restricted to that domain.

(5.5) $\mathcal{D}$ <u>is an open set</u>.

For every point of $\mathcal{D}$ has a connected neighborhood N which is in Ω and hence in $\mathcal{D}$. Hence $\mathcal{D}$ is open.

6. <u>Extension of the solution</u>. <u>Trajectories</u>. (6.1) The existence theorem yields a solution $\vec{x}(t) = \vec{x}^0$, and valid over a certain interval I_τ containing t^0, and such that $\vec{x}(t^0)$. Let t'^0 be any point of I_τ with $\vec{x}(t'^0) = \vec{x}'^0$. There is a similar solution $\vec{x}'(t)$ valid over an interval $I_{\tau'} \ni t'^0$ and such that $\vec{x}'(t'^0) = \vec{x}'^0$. Moreover by the

uniqueness property $\vec{x}(t) = \vec{x}'(t)$ for $t \in I_\tau \cap I_{\tau'}$. We call $\vec{x}'(t)$ the <u>continuation</u> of $\vec{x}(t)$ to $I_{\tau'}$ and we now define $\vec{x}(t)$ on $I_{\tau''} = I_\tau \cup I_{\tau'}$ by assigning to it the values $\vec{x}'(t)$ for t in $I_{\tau'} - I_\tau$. We thus extend the definition of $\vec{x}(t)$ to $I_{\tau''}$. By repetition of this process we thus arrive at a maximal interval $t' < t < t''$ over which $\vec{x}(t)$ is defined. It may be of course that $t' = -\infty$ or $t'' = +\infty$ or both. The analogy with the classical process of analytical continuation is obvious.

(6.2) If we start from $(\vec{x}^0, t^0) \in \mathcal{D}$, the set of points $(\vec{x}, t)$ reached is such that from $(\vec{x}^0, t^0)$ to any one of them there may be constructed a chain of boxes $B_1, \ldots, B_r$ each in a domain such as $\mathcal{D}$ and with B_i, B_{i+1} overlapping. Since the boxes are connected so is their union and since $\mathcal{D}$ is a component of all the sets $\mathcal{D}$, every $B_i \subset \mathcal{D}$, hence $(\vec{x}, t) \in \mathcal{D}$. <u>Thus by continuation we never leave</u> $\mathcal{D}$.

(6.3) The set of points $(\vec{x}(t), t)$, $t \in I$, is called the <u>trajectory defined by</u> $\vec{x}(t)$. We shall also say: the <u>trajectory</u> $\vec{x}(t)$, rather than the solution $\vec{x}(t)$.

(6.4) <u>A trajectory is an arc.</u>

Denote the trajectory of (6.2) by Γ, and consider the transformation $\phi : I_\theta \longrightarrow \Gamma$ defined by $\phi(t) = (\vec{x}(t), t)$, $t \in I_\theta$. It is clear that ϕ is continuous, univalent, and $\phi I_\theta = \Gamma$, Hence ϕ is one-one and onto. It follows already that if I is any closed subinterval of I_θ, (hence $\bar{I}$ is compact), the values of ϕ on $\bar{I}$ give rise to a topological mapping ψ of $\bar{I}$ onto a subset J of Γ. There remains to prove ϕ^{-1} continuous or ϕ open. Consider $(\vec{x}^0, t^0) \in \Gamma$ with $\phi^{-1}(\vec{x}^0, t^0) = t^0 \in I_\theta$. Let 2τ be a lower bound for the distance from t^0 to the end points of I_θ. Then if $U = \Gamma \cap \sum(\vec{x}^0, t^0, \tau)$, the set $\phi^{-1}U$ is contained in an interval I such that $\bar{I} \subset I_\theta$. Since ψ is topological, $\phi^{-1}U$ is an open set of $\bar{I}$ and since it is in I, it is an open set of I and hence of I_θ. Therefore ϕ is open, hence it is topological and Γ is an arc.

§3. CONTINUITY PROPERTIES

7. If $(\vec{x}^0, t^0) \in \mathcal{D}$ then we may enclose it in a box. $B(\alpha, \tau)$. It will be recalled that for any $t, t^0 \in I_\tau$ and $\vec{x}^0 \in U_\alpha$ there was obtained a solution

$$\vec{x}(t) = \vec{x}^0 + (\vec{x}^1 - \vec{x}^0) + \dots \tag{7.1}$$

where under the conditions just stated the series is uniformly convergent. Let us consider $\vec{x}(t)$ as a function of t, $\vec{x}^0$, t^0, and write it accordingly $\vec{x}(t, \vec{x}^0, t^0)$. Then

(7.2) *If t, $t^0 \in I_\tau$ and $\vec{x}^0 \in U_\alpha$, then $\vec{x}(t, \vec{x}^0, t^0)$ is a continuous function of $(t, \vec{x}^0, t^0)$.*

Owing to the uniform convergence it is only necessary to prove that $\vec{x}^{m+1} - \vec{x}^m$ is continuous over the range under consideration, or in the last analysis that $\vec{x}^m$ has this property. This is trivial for $\vec{x}^0$ so we assume it for $\vec{x}^m$ and prove it for $\vec{x}^{m+1}$. In the proof of the existence theorem it has been shown that under the conditions under consideration all the $\vec{x}^m$ are confined in a certain box $B' \subset \mathcal{D}$ in whose points $\vec{p}(\vec{x}, t)$ is continuous and bounded. It follows that in $(4.2)_m$ the integrand is continuous and bounded under our conditions. Since the integral is thus a continuous function of $(t, \vec{x}^0, t^0)$, the same holds for $\vec{x}^{m+1}$ and (7.2) follows.

We now remove the restrictions on t, $\vec{x}^0$, t^0, save of course that only points $(\vec{x}, t) \in \mathcal{D}$ are to be considered and we will prove continuity under the same conditions. More precisely:

(7.3) *If $(\vec{x}^0, t^0) \in \mathcal{D}$ and $\vec{x}(t, \vec{x}^0, t^0)$ is the solution such that $\vec{x}(t^0, \vec{x}^0, t^0) = \vec{x}^0$ then $\vec{x}(t, \vec{x}^0, t^0)$ is continuous in $(t, \vec{x}^0, t^0)$.*

Consider the trajectory Γ_0 through $(\vec{x}^0, t^0)$. Combining successive approximations with the process of continuation we find that the arc of Γ_0 from $(\vec{x}^0, t^0)$ to $(\vec{x}, t)$, being a compact set, may be covered by a finite number of boxes such as B, say $B_1, \dots, B_r$, where the first contains

$(\vec{x}^0,t^0)$, the last $(\vec{x},t)$, and consecutive boxes overlap. More precisely if $B_i = I_{\tau_i} \times U_i$ then I_{τ_i}, $I_{\tau_{i+1}}$ overlap and so we may choose a value t^1 in their common part. We will now keep $(\vec{x}^0,t^0)$ fixed and consider a new variable initial point $(\vec{x}',t')$ in B_1. If Γ is the trajectory through this point then it meets the set $t = t^1$ at a certain point $(\vec{x}^1,t^1)$ whose coordinate $\vec{x}^1$ is a continuous function of $(\vec{x}',t')$ by (7.2). It follows that we can choose a sphere $\sum(\vec{x}^0,t^0,\rho) \subset B_1$, with ρ so small that when $(\vec{x}',t') \in \sum$ then $(\vec{x}^1,t^1) \in B_2$. Now for $(\vec{x}^1,t^1) \in B_2$ and $t \in I_{\tau_2}$ the solution $\vec{x}(t)$ such that $\vec{x}(t^1) = \vec{x}^1$, is a continuous function of $\vec{x}',t'$ and hence of $(t,\vec{x}',t')$ for $t \in I_{\tau_2}$ and $(\vec{x}',t') \in \sum$. Confining now $(\vec{x}',t')$ to $\sum$, Γ will contain the point $(\vec{x}^1,t^1) \in B_2$, and so it will contain a point $(\vec{x}^2,t^2)$, where $\vec{x}^2 = \vec{x}(t^2)$. The same continuity argument shows that when ρ is sufficiently small $(\vec{x}^2,t^2) \in B_3$ and we may now proceed as before. After a finite number of steps we shall find a ρ so small that for $(\vec{x}',t') \in \sum$, $\vec{x}(t,\vec{x}',t')$ will remain in B_r and be continuous in $(t,\vec{x}',t')$ for $t \in I_{\tau_r}$ and $(\vec{x}',t) \in \sum$. This implies (7.3).

(7.4) Several noteworthy consequences may be drawn from the preceding results. The notations being the same corresponding to $M(\vec{x}^0,t^0)$ we may choose ε, δ such that the sets $t^0 \times \mathscr{S}(\vec{x}^0,\varepsilon)$, $\sum(M,\delta) \subset B_1$. Consider now $\vec{x}(t,\vec{x}',t')$, $(\vec{x}',t') \in \sum$. By the existence theorem the function is defined for every $t \in I_{\tau_1}$ and hence for $t = t^0$. Let $\vec{x}(t,\vec{x}',t') = \vec{x}''(t)$, so $\vec{x}(t^0,\vec{x}',t') = \vec{x}''(t^0)$. By (7.3) $\vec{x}''(t)$ is a continuous function of $(\vec{x}',t')$. Therefore δ may be chosen such that $(\vec{x}''(t^0),t^0) \in t^0 \times \mathscr{S}(\vec{x}^0,\varepsilon)$. In other words any trajectory passing near enough to $(\vec{x}^0,t^0)$ in $\mathcal{D}$ will cross $t^0 \times \mathscr{S}(\vec{x}^0,\varepsilon)$: and of course the converse will hold if $\delta \geq \varepsilon$. We may thus say that the trajectories passing close to a given point $M(\vec{x}^0,t^0) \in \mathcal{D}$ at time t^0 include all those passing quite close to M in $\mathcal{D}$. Or explicitly:

(7.5) *Corresponding to any* $M \in \mathcal{D}$ *and any* $\varepsilon > 0$ *there exists a* $\delta > 0$ *such that a trajectory crossing* $\Sigma(M,\delta)$ *intersects the hyperplane* $t = t^0$ *within* $\mathcal{S}(\vec{x}^0,\varepsilon)$ (i.e., *it meets* $t^0 \times \mathcal{S}(\vec{x}^0,\varepsilon)$).

(7.6) An extension of the continuity theorem in a new direction is the following. Suppose $\vec{p}$ is in fact a function $\vec{p}(\vec{x},t,\vec{y})$ which is continuous and bounded in a region Ω of $\mathcal{W} \times \mathcal{U}_y$. We suppose moreover that with respect to Ω the Lipschitz condition (4.1.3) is still fulfilled in the same form as before. The *domain of continuity* Δ is defined as before (5.4) as a component of the union of all the sets Ω. We have now the following stronger result which extends (7.3) and may understandably be stated in the brief form:

(7.7) *The solution is continuous in* $(t,\vec{x}^0,t^0,\vec{y})$ *when* $(\vec{x}^0,t^0,\vec{y})$ *ranges over* Δ.

The proof may be related to (7.3) by a well known device. Enlarge (3.2) by adding the differential equation

$$\frac{d\vec{y}}{dt} = 0 \tag{7.8}$$

so that (3.2) and (7.8) form a system such as (3.2) for the vector $(\vec{x},\vec{y})$. Owing to the special form of (7.8) the Lipschitz condition(4.1.3) still suffices to prove the existence theorem and hence all its corollaries, including among them (7.3), which in the present instance becomes (7.7)

(7.9) Consider again the trajectory Γ_0 and an arc MN of Γ_0, where $(\vec{x}^0,t^0)$ are the coordinates of M and $(\vec{x}^1,t^1)$ are those of N. Introduce the two closed sets $S_0 = t^0 \times \bar{\mathcal{S}}(\vec{x}^0,\rho_0)$, $S_1 = t^1 \times \bar{\mathcal{S}}(\vec{x}^0,\rho_1)$, where ρ_0, ρ_1 are so chosen that the two sets are in Ω. Let $M' \in S_0$ and let $\Gamma' : \vec{x}(t,M')$ be the trajectory through M'. It is clear from the argument proving (7.3) that for ρ_0 small enough $\vec{x}(t,M')$ may be extended throughout the whole closed interval $\bar{I} : t^0 \leq t \leq t^1$. Consider a mapping ϕ of the compact product $\bar{I} \times S_0$ into $\mathcal{D}$ sending $\bar{I} \times M$ into MN and

$\overline{I} \times M'$ into $M'N'$. This mapping ϕ is <u>univalent</u>, since $(\vec{x}', t')$ and $(\vec{x}'', t'')$ certainly have different images if $t' \neq t''$ or $\vec{x}' \neq \vec{x}''$, (the latter since otherwise distinct trajectories would meet). Since $\overline{I} \times S_0$ is compact ϕ is topological.

Take now a small n-cell E_0^n containing M and contained in S_0. Applying ϕ merely to E_0^n or to $\overline{I} \times E_0^n$ and denoting the hyperplanes $t = t^i$ by π_i we find:

(7.10) <u>Given a sufficiently small n-cell E_0^n in π_0 containing M, a trajectory passing through a point M' of E_0^n intersects π_1 in a single point N' such that $M' \longrightarrow N'$ defines a topological mapping θ of E_0^n onto a similar $E_1^n \subset \pi_1$, containing N and of course $N = \theta M$.</u>

(7.11) <u>The circumstances remaining the same and E_0^n being small enough, let λ be the arc M'N' of the trajectory through M', and let $\lambda(t)$ be the point of λ corresponding to any $t \in \overline{I}$, $\overline{I} : t^0 \leq t \leq t^1$. Then $(M', t) \longrightarrow \lambda(t)$ defines a topological mapping ϕ of the cylinder $\overline{I} \times E_0^n$ such that $\phi(\overline{I} \times M) = MN$, $\phi(\overline{I} \times M') = \lambda$.</u>

This last result embodies essentially the so-called "field" theorem for minimizing arcs in the Calculus of Variations.

(7.12) <u>General solution</u>. This concept may now be introduced with reasonable clarity. Let $\mathcal{V}_c$ be n-dimensional and $\vec{f}(t, \vec{c})$ a function such that: (a) for $\vec{c}$ in a certain region Λ of $\mathcal{V}_c$, $\vec{f}$ is a solution of (3.2) in the domain $\mathcal{D}$; (b) if $M_0 = \vec{f}(t^0, \vec{c}^0)$, $M = \vec{f}(t, \vec{c})$, where $\vec{c}$, $\vec{c}^0$ are in Λ, there are n-cells E^n in $\mathcal{V}_x \times t^0$ and $\mathcal{E}^n$ in Λ, such that for $\vec{c} \in \Lambda$ the correspondence $\vec{c} \longrightarrow M$ is a topological mapping ϕ of $\mathcal{E}^n$ onto E^n, such that $\phi\vec{c}^0 = M_0$. In other words $\vec{c}$ may be chosen in $\mathcal{E}^n$ so as to yield any solu- with its initial value at t^0 in E^n. The function $\vec{f}(t, \vec{c})$ is known as a <u>general</u> solution. By (7.10) the concept of general solution is manifestly independent of the particular point M_0 chosen on the trajectory $\vec{f}(t, \vec{c}^0)$, i.e. it does not depend upon t^0 but solely upon the trajectory itself.

(7.13) To avoid undue repetitions we may state here and now that in dealing with complex functions the definition of general solution will be the same save that $\mathcal{V}_c$ will be complex n-dimensional, the cells 2n-dimensional and the mapping η analytical. An explicit reformulation will not be necessary.

§4. ANALYTICITY PROPERTIES

8. (8.1) The argument in deriving the continuity properties of $\vec{x}(t)$ rests essentially upon the following three propositions:

(a) A continuous function of a continuous function is continuous.

(b) If $\vec{f}(\vec{x},t)$ is continuous in $(\vec{x},t)$ over a suitable range so is $\int_{t_0}^{t} \vec{f}(\vec{x},t)dt$.

(c) A uniformly convergent series of continuous functions is continuous.

These three propositions made it possible to "transfer" continuity from the approximations $\vec{x}^m(t)$ to the solution $\vec{x}(t)$ itself. Since these three properties hold also with "continuous" replaced by "analytic" or "holomorphic" the same transfer will operate for analyticity or holomorphism. It will be necessary however to distinguish carefully between analyticity as to $\vec{x}^0$, as to t, t^0, or as to all three. The difference arises from the fact that we may wish to consider not only $\vec{p}(\vec{x},t)$ analytic in $(\vec{x},t)$ but also merely in $\vec{x}$ alone, or in certain additional parameters that may be present in $\vec{p}$.

(8.2) Suppose $\vec{p}(\vec{x},t)$ in (3.2) analytic in $\vec{x}$ and continuous in $(\vec{x},t)$ in a certain region Ω. If $(\vec{\xi},\eta) \in \Omega$ we may choose a box $B(2\alpha,2\tau)$ of center $(\vec{\xi},\eta)$ contained in Ω. As a consequence the p_i and their partial derivatives $\frac{\partial p_i}{\partial x_j}$ are continuous on the compact set $\bar{B}$, and so bounded in B. Hence (4.1.1), (4.1.2) hold in B (5.2) and the existence theorem is applicable in B.

(8.3) We now proceed as before replacing continuity by analyticity wherever need be and obtain first an analogue Δ of $\mathcal{O}$ which we call in any case domain of analyticity, then we have the following properties, which we merely state since the proofs are unmodified. The space $\mathcal{V}_x$ and the values of the variables and functions may be real or complex unless otherwise restricted.

(8.4) If $\vec{p}(\vec{x},t)$ is analytic in both variables and Δ is the domain of analyticity then the solution $\vec{x}(t,\vec{x}^0,t^0)$ such that $(\vec{x}(t),t) \in \Delta$, $\vec{x}(t^0,\vec{x}^0,t^0) = \vec{x}^0$ is analytic in all three arguments. (see (7.3).

(8.5) If $\vec{p}(\vec{x},t)$ is merely continuous in t (t real) then $\vec{x}(t,\vec{x}^0,t^0)$ is merely analytic in $\vec{x}^0$. (see 7.3).

(8.6) If $\vec{p}(\vec{x},t,\vec{y})$ is analytic in $\vec{y}$ also and Δ is defined accordingly as in (7.6) then $\vec{x}(t,\vec{x}^0,t^0,\vec{y})$ is analytic in all arguments in the case (8.4) and in $(\vec{x},\vec{y})$ alone in the case (8.5). (see 7.7)

9. (9.1) Poincaré's expansion theorem. Let the differential equation with t real and the domain of analyticity Δ:

$$\frac{d\vec{x}}{dt} = \vec{p}(\vec{x},t,\vec{y}) \tag{9.1.1}$$

possess for $\vec{y} = 0$ a solution $\vec{\xi}(t)$ on the closed interval $\bar{I} : t^0 \leq t \leq t^1$ such that the $p_i(\vec{x},t,\vec{y})$ may be expanded in power series of the $(x_i - \xi_i(t))$ and the y_j convergent in some range

$$|x_i - \xi_i(t)| < \alpha, \quad |y_j| < \alpha \tag{9.1.2}$$

$$i = 1, 2, \ldots, n; \quad j = 1, 2, \ldots, r,$$

$(r = \dim \mathcal{V}_y)$ and this for every $t \in \bar{I}$. Then setting $\vec{\xi}(t^0) = \vec{\xi}^0$, the equation (9.1.1) has a solution $\vec{\xi}(t,\vec{x}^0,\vec{y})$ such that $\vec{\xi}(t,\vec{\xi}^0,0) = \vec{\xi}(t)$ and that the $\xi_i(t,\vec{x}^0,\vec{y})$ may be expanded in power series of the $(x_i^0 - \xi_i^0)$ and the y_j convergent for $t \in \bar{I}$ and $(\vec{x}^0,\vec{y})$ in a certain toroid

(9.1.3) $$|x_i - \xi_i^0| < \alpha, \quad |y_j| < \alpha.$$

In particular if $\vec{\xi}(t,\vec{\xi}^0,\vec{y}) = \vec{\xi}(t,\vec{y})$ then the latter is a solution such that $\vec{\xi}(t,0) = \vec{\xi}(t)$ and that $\vec{\xi}(t,\vec{y})$ may be expanded in power series of the y_j convergent in a toroid

(9.1.4) $$|y_j| < \alpha.$$

This theorem has been utilized by Poincaré, Picard and others in many fruitful ways in questions of approximations, likewise in the search for solutions with special properties (periodicity among others) neighboring specified solutions.

If we make the change of variable $\vec{x} - \vec{\xi}(t) = \vec{z}(t)$, then $\vec{z}(t)$ satisfies the differential equation

(9.2) $$\frac{d\vec{z}}{dt} = \vec{q}(\vec{z},t,\vec{y})$$

which behaves like (9.1.1) save that now $\vec{\xi}(t)$ is replaced by 0 and that q_i may be expanded in a power series in the z_i, y_j in the toroid $\mathcal{J}(\alpha)$: $|z_i| < \alpha$, $|y_j| < \alpha$. $(i = 1, 2, \ldots, n;\ j = 1, 2, \ldots, r)$. We merely need to prove now for (9.2) the analogue of (9.1) with $\vec{\xi}(t) = \vec{\xi}^0 = 0$, and the existence of a solution $\vec{\zeta}(t,\vec{z}^0,\vec{y})$ possessing a series expansion in powers of the z_i^0, y_j valid for $t \in \bar{I}$ and $(\vec{z}^0,\vec{y})$ in a certain toroid $\mathcal{J}(\beta)$.

Let (9.2) be amplified by

(9.3) $$\frac{d\vec{y}}{dt} = 0.$$

Thus (9.2), (9.3) form a system in the unknown $(\vec{y},\vec{z})$ with the right hand sides holomorphic in $\mathcal{J}(\alpha)$ for every $t \in \bar{I}$. The associated domain of analyticity Δ_1 in $\mathcal{V}_y \times \mathcal{V}_z \times L$ contains the point $(0,0,t^1)$ and the product $\mathcal{J}(\alpha) \times \bar{I}$. By the mechanism of the proof of (7.3) as paraphrased for (8.5) there is a solution $\vec{\zeta}(t,\vec{y}^0.\vec{z}^0)$ such that $\vec{\zeta}(t,0,0) = 0$, $\vec{y} = \vec{y}^0$ valid for every $t \in \bar{I}$ and analytic in the initial values $\vec{y}^0$, $\vec{z}^0$ when they are in a certain box B of center

(0,0). Hence the solution $\vec{\xi}(t,\vec{y}^0\vec{z}^0)$ may be expanded in a power series of the y_i^0, z_i^0 valid in a toroid $\mathcal{J}(\beta) \subset B$. Since $\vec{y} = \vec{y}^0$, this is the required property of $\vec{\xi}$ and (9.1) is proved.

§5. EQUATIONS OF HIGHER ORDER

10. We refer particularly to the single equation (1.4.1). Since it is equivalent to (1.4.2) the extension of the preceding results is immediate. It may be observed that the Lipschitz condition is equivalent here to:

$$\left\|\left(a_1(x,t)\frac{d^{n-1}x}{dt^{n-1}} + \dots + a_n(x,t)\right) - \left(a_1(x',t)\frac{d^{n-1}x'}{dt^{n-1}} + \dots\right)\right\|$$

$$(10.1) \qquad < k\sum \dots .$$

and this is the form in which it is to be utilized in connection with the existence theorem. With this interpretation of the Lipschitz condition we may state the following propositions which follow from those already proved.

(10.2) <u>The symbols and conditions being as in</u> (4.1) <u>there exists a unique solution</u> $x(t)$ <u>of</u> (1.4.1) <u>valid in a certain interval</u> $|t-t^0| < \tau$ <u>and such that</u>

$$(10.2.1) \qquad \left(\frac{d^p x}{dt^p}\right)_{t = t_0} = x^0_{p+1}, \qquad p \leq n-1.$$

<u>This solution is a continuous function of</u> t, t^0 <u>and the</u> x^0_{p+1} <u>after the manner of</u> (7.3). <u>Moreover if the coefficients</u> a_i <u>are analytic under conditions similar to those of</u> (8.4) <u>then the solution has the appropriate analyticity properties</u>.

§6. SYSTEMS IN WHICH THE TIME DOES NOT FIGURE EXPLICITLY

11. Systems. They are the systems of the form

$$\frac{d\vec{x}}{dt} = \vec{p}(\vec{x}). \tag{11.1}$$

They arise for instance in dynamics whenever the applied forces depend solely upon the coordinates and their derivatives but not upon the time.

(11.2) If L denotes the space of t (real line) and S is a region of $\mathcal{V}_x$ in which the partial derivatives $\frac{\partial p_i}{\partial x_j}$ exist and are continuous and bounded, then by (5.2) we may take $\Omega = L \times S$, in the proof of the theorem of existence. This is a decidedly more restricted region than in the general case, but it will remove certain difficulties that would be caused by the Lipschitz condition in connection with the important considerations of (12).

(11.3) Let $\mathcal{D}$ be a component of the set of all the points of $\mathcal{V}_x$ which have a neighborhood such as S. The domain of continuity is now by definition a set $L \times \mathcal{D}$. However the significant set is actually $\mathcal{D}$ and it will be called the characteristic domain of continuity of (11.1), or merely the domain wherever the meaning is otherwise clear. A similar characteristic domain of analyticity $\Delta \subset \mathcal{D}$ may be introduced when $\vec{p}$ is analytical.

(11.4) Beyond this point the simplifications in the various corollaries to (4.1) offer no difficulties and are left to the reader.

(11.5) Suppose that $\vec{x}(t)$ is a solution of (11.1). If we make the change of variables $t' = t - \tau$, the form of (11.1) is unchanged. Hence $\vec{x}(t-\tau)$ is a solution whatever τ. Thus from a single solution, we may derive here a whole family of solutions depending upon an arbitrary parameter τ.

(11.6) Consider the trajectory Γ_τ corresponding to $\vec{x}(t-\tau)$, in the space $L \times \mathcal{V}_x$ and let γ be its projection in $\mathcal{V}_x$. Γ_τ is the set of points $(\vec{x}(t-\tau),t) \in L \times \mathcal{V}_x$, and so γ is the set of points $\vec{x}(t-\tau)$ in $\mathcal{V}_x$. With Poincaré we will call γ a characteristic of (11.1). This characteristic is independent of τ. Suppose in fact that $\Gamma_{\tau'}$ has the projection γ'. Thus γ' is the set of all points $\vec{x}(t-\tau')$. Now if M_t, M'_t are the points of γ, γ' corresponding to the value t, evidently $M_{t+\tau-\tau'} = M'_t$ and so $\gamma = \gamma'$. In other words γ, γ' are merely different parametrizations of the same locus.

The characteristics are all contained in the domain $\mathcal{D}$ and in this domain through each point $M(x^0)$ there passes a single characteristic γ. For the characteristics are the projections of the trajectories Γ_τ through the points $(\vec{x}^0,\tau)$ for all τ. If $\vec{x}(t)$ is the trajectory through $(\vec{x}^0,0)$, i.e. the solution such that $\vec{x}(0) = \vec{x}^0$, then $\Gamma_\tau : \vec{x}(t-\tau)$ is a trajectory and hence the trajectory through $(\vec{x}^0,\tau)$ (since it is unique). Thus all the Γ_τ are merely the trajectories corresponding to all the solutions $\vec{x}(t-\tau)$, and as we have seen the corresponding characteristic is unique.

(11.7) Critical points. A critical point of (11.1) is a point (= a vector) $\vec{x}^0$ of the domain $\mathcal{D}$ which annuls $\vec{p}(\vec{x})$, i.e. a solution of $\vec{p}(\vec{x}) = 0$ in $\mathcal{D}$. Evidently $\vec{x}(t) = \vec{x}^0$ is a characteristic (sometimes called a "point-characteristic"). The corresponding trajectory is unique and is the parallel to the t-axis consisting of all the points (x^0,t). Unless otherwise stated the term "characteristic" will be reserved in the sequel for those which are not critical points.

(11.8) If $\vec{x}^0$ is a critical point then no trajectory other than $\vec{x}(t) = \vec{x}^0$ may reach $(\vec{x}^0,t^0)$ by the process of extension.

Since $(\vec{x}^0,t^0) \in L \times \mathcal{V}_x$ if a second trajectory $\vec{x}^*(t)$ could reach it there would be two trajectories through the point, which is ruled out.

As a corollary:

(11.9) <u>If</u> $\vec{x}(t)$ <u>tends to a critical point</u> $\vec{x}^0$, <u>then</u> $t \to +\infty$ <u>or</u> $t \to -\infty$.

12. (12.1) Side by side with the system (11.1) we may consider the system

$$(12.2) \qquad \frac{dx_1}{p_1(x_1,\ldots,x_n)} = \cdots = \frac{dx_n}{p_n(x_1,\ldots,x_n)}$$

which may be interpreted in the following way. We first recall that a <u>differentiable curve</u> (one dimensional differentiable manifold) in $\mathcal{U}_x$ is a locus λ such that if $M \in \lambda$ then a certain neighborhood μ of M in λ admits a parametric representation $x_i = f_i(u)$, or equivalently $\vec{x} = \vec{f}(u)$, $a < u < b$, where the f_i are univalent and differentiable on the interval. Let us agree to consider likewise a single point as a differentiable curve. We may then say that (12.2) requires to find differentiable curves λ such that at each point of λ the vector $d\vec{x}$ is collinear with $\vec{p}(\vec{x})$, or is zero wherever $\vec{p}(\vec{x}) = 0$.

It is clear that the characteristic γ through any point M of $\mathcal{D}$ is a suitable λ.

Suppose first M critical. Then γ and λ both reduce to M and so they coincide. Suppose now M non-critical. At the point M one of the components of $\vec{p}$, say $p_h \neq 0$. Since p_h is continuous and $M \in \mathcal{D}$ there is a sphere $\mathcal{S}(M,\rho)$ in $\mathcal{D}$ such that $p_h \neq 0$ and is bounded away from zero in the sphere. As a consequence, in the sphere the system (12.2) is equivalent to

$$(12.3.h') \qquad \frac{dx_i}{dx_h} = q_i(\vec{x}) = \frac{p_i(\vec{x})}{p_h(\vec{x})}, \qquad i \neq h.$$

This system is of the form (3.1) with x_h in place of t. Owing to our assumptions (see 11.2) the $\frac{\partial q_i}{\partial x_j}$ exist and are continuous and bounded in $\mathcal{S}(M,\rho)$ and so (4.1) may be applied within the sphere and with x_h as the independent variable. It implies that there is a unique differentiable arc μ in $\mathcal{S}(M,\rho)$ containing M and along which (12.3.h) holds, i.e. along which (12.2) holds. Now the arc μ' of the characteristic γ through M, situated in $\mathcal{S}(M,\rho)$ is a differentiable arc such as μ, and therefore $\mu' = \mu$. Thus

(12.4) <u>The solution of</u> (12.2) <u>at any point</u> M <u>of</u> $\mathcal{O}$ <u>is the characteristic through</u> M.

13. <u>Closed characteristics</u>. A characteristic γ is said to be closed whenever it is a Jordan curve (simple closed curve) at a positive distance from the set of critical points. Thus γ is compact and every point of γ is reached from a given point M in a finite time. It follows that the solution $\vec{x}(t)$ corresponding to γ is periodic. Conversely if $\vec{x}(t)$ is periodic then its characteristic γ is closed. For γ is compact and all its points are reached in finite time, hence γ is at a positive distance from the set of critical points. Let τ be the period. Then $\vec{x}(0) = \vec{x}(\tau)$ and $\vec{x}(0) \neq \vec{x}(\theta)$, $0 < \theta < \tau$, else θ would be the period and not τ. Since as t varies from 0 to τ, limits excluded, $\vec{x}(t)$ describes an arc of γ, γ is a Jordan curve. Thus

(13.1) <u>A n.a.s.c. for a characteristic</u> γ <u>to be closed is that its solution</u> $\vec{x}(t)$ <u>be periodic</u>.

14. We shall now prove for characteristics certain results in close analogy with those of (7.10, 7.11) for trajectories. The basic properties are (14.4, 14.6, 14.8). The first two state roughly that an arc of characteristic may be imbedded in a "tube" of characteristics. This property is very convenient in the topological applications. The difficulties in the proof arise from the presence of closed characteristics. To simplify matters it will be assumed henceforth that <u>the basic system</u> (11.1) <u>is analytical</u>.

(14.1) Let M be a non-critical point, γ the characteristic through M, E^{n-1} an elementary analytical (n-1)-cell through M, represented by

(14.1.1) $$x_i = \phi_i(u_1,\ldots,u_{n-1}), \qquad \|\vec{u}\| < \alpha$$

or in vector notation

(14.1.2) $$\vec{x} = \vec{\Phi}(\vec{u}),$$

where $\vec{\Phi}(0) = \vec{x}^0$, and the Jacobian matrix of the functions ϕ_i is of rank (n-1) on E^{n-1}. The latter is said to be <u>transverse</u> to γ at M whenever γ is not tangent to E^{n-1} at M. This means in substance that the differential vector $d\vec{x}$ along γ at M is linearly independent of those along E^{n-1} at M. In other words its components are <u>not</u> of the form

(14.1.3) $$dx_i = \sum \frac{\partial x_i}{\partial u_j} du_j$$

or again the vector $\vec{p}(\vec{x}^0)$ is independent of the vectors represented by (14.1.3) for $\vec{u} = 0$. In the last analysis this means that the determinant

$$D = \left| \frac{\partial x_i}{\partial u_1}, \ldots, \frac{\partial x_i}{\partial u_{n-1}}, p_i(\vec{x}) \right| \neq 0$$

for $\vec{x} = \vec{x}^0$, $\vec{u} = 0$. Owing to the continuity of D, it will not vanish for $\|\vec{u}\|$ and $\|\vec{x}-\vec{x}^0\|$ small. Hence if we choose α in (14.1.1) sufficiently small, then the characteristic γ' through any point M' of E^{n-1} is likewise not tangent to the cell at M'. That is to say under our assumption the cell E^{n-1} is transverse to all the characteristics which cross it.

(14.2) The notations remaining the same let us associate explicitly with γ' the solution $\vec{x}(t,M')$ such that $\vec{x}(0,M') = \vec{x}'^0 = M'$. This solution satisfies the identical relation

(14.2.1) $$\vec{x}(t) = \vec{\Phi}(\vec{u}) + \int_0^t \vec{p}(\vec{x})\, dt.$$

or explicitly

$$(14.2.2)\quad x_i(t) = \phi_i(u_1,\ldots,u_{n-1}) + \int_0^t p_i(x_1,\ldots,x_n)\,dt.$$

Consider this system as a system of n equations in the n unknowns $u_1, \ldots, u_{n-1}$ t. It is an analytical system and for the values zero of all the unknowns the Jacobian determinant of the right hand sides is precisely equal to the value of D for $\vec{u} = 0$, $\vec{x} = \vec{x}^0$, and hence $\neq 0$. Therefore (14.2.2) is a regular transformation and defines a topological mapping f of a suitable set: $\|\vec{u}\| \leq \beta$, $|t| \leq \tau$ on $\mathcal{V}_x$. Choose now $\alpha < \beta$ and let I_τ denote the time interval $|t| < \tau$. Thus f will map topologically the cylinder $E^{n-1} \times I_\tau$ on $\mathcal{V}_x$ in such manner that (a) $fM' = M'$ and in particular $fM = M$; (b) $f(M' \times I_\tau)$ is mapped into the arc of γ' described in the time $-\tau < t < \tau$.

It is convenient to introduce here the following "cylindrical" terminology. The image $C = f(E^{n-1} \times I_\tau)$ will be referred to as an <u>open characteristic cylinder</u> and the arcs $f(M' \times I_\tau)$, $f(M \times I_\tau)$ as the <u>generators</u> and the <u>axis</u> of the cylinder. When the dimension is two we shall say "characteristic rectangle". We note the following two properties:

(14.3) <u>Every characteristic passing sufficiently near to M must have an arc in the characteristic cylinder C and hence it must cross the cell E^{n-1} transverse to γ at M.</u>

For f is topological and hence C is a neighborhood of M. Consequently if the characteristic γ' passes sufficiently near to M it will meet C in some point P and therefore γ' will contain the generator through P.

(14.4) <u>If γ' through $M' \in E^{n-1}$ followed beyond M' returns to a point M" of E^{n-1} then it must take a time at least τ to do so.</u>

For f maps $M' \times I_\tau$ topologically on $\mathcal{V}_x$, and to return to E^{n-1} it must take at least the time τ that it takes to describe the generator through M' beyond M'.

(14.5) The situation remaing the same let γ be followed beyond M up to a point N $(\vec{x}^1)$ and let E_1^{n-1} be transverse to γ at N and so small that it does not meet E^{n-1}. Let the cell have the parametric representation similar to (14.1.2):

$$(14.5.1) \qquad \vec{x} = \vec{\psi}(\vec{v}), \qquad \|\vec{v}\| < \beta, \ \vec{\psi}(0) = \vec{x}'.$$

Let t_1 be the time at which N is reached along γ beyond M. That is to say $N = \vec{x}(t^1, M)$. As a consequence of (7.10) if γ' is followed beyond M' during a time t_1 then a point will be reached which may be made arbitrarily close to N by choosing M' sufficiently close to M. Combining this with (14.3) we find that under the circumstances γ' will intersect E_1^{n-1}. Let N' denote its first intersection with E_1^{n-1} beyond M'. If the constant α in (14.1.1) is sufficiently small then this will occur for every point M' of E^{n-1} and even of its closure $\bar{E}^{n-1}$. This will be assumed henceforth. We shall in fact actually suppose α so small that the representation (14.1.1) is valid throughout $\bar{E}^{n-1}$.

Consider now a segment $\Lambda : 0 \leq s \leq 1$ and the transformation $F : E^{n-1} \times \Lambda \longrightarrow \mathcal{U}_x$ defined in the following manner. Let the arc M'N' of γ' be described in time t_1'. Then F sends $M' \times \Lambda$ into M'N' in such manner that the point $M' \times s$ is imaged into the point of γ' reached at time st_1'. In particular $F(M' \times 0) = M'$, $F(M' \times 1) = N'$. It is clear also that F is continuous, i.e. it is a <u>mapping</u> in the technical sense. We shall show that if the constant α is sufficiently small then F is also <u>univalent</u> i.e. that it images distinct points into distinct points.

Suppose in fact that F images two distinct points $(M' \times s')$ and $(M'' \times s'')$ into the same point P. Thus P is on both M'N' and M"N". This can only arise if γ' followed say beyond M' passes through M". Now we may assume α so small that (14.4) becomes applicable. Thus it will require a time at least τ to describe the arc $M'M''$ of γ'. Moreover if F fails to be univalent however small one chooses α, the preceding situation must arise for some M' arbitrarily near M. Hence by an obvious limiting process γ must return to M after a time t_2 where $\tau \leq t_2 \leq t_1$. Since MN is an arc this is impossible. Hence F is univalent.

Thus F is a one-one mapping of the compact set $\overline{E}^{n-1} \times \Lambda$ onto a subset C of $\mathcal{U}_x$. As a consequence F is <u>topological</u>. The image set C of the closed cylinder $\overline{E}^{n-1} \times \Lambda$ is called a <u>closed characteristic cylinder</u>. The arc MN is referred to as the <u>axis</u> of the cylinder, the transverse cells E^{n-1}, E_1^{n-1} as its <u>bases</u>, the images $F(M' \times \Lambda)$ as its <u>generators</u>. We have therefore proved:

(14.6) <u>Let γ be any characteristic, MN an arc of γ, E^{n-1} and E_1^{n-1} elementary analytical cells transverse to γ at M and N. Then there exist a closed characteristic cylinder whose axis is MN and whose bases are in the two given cells.</u>

(14.6.1) When n = 2 the cylinder is called a rectangle and the cells are arcs but otherwise nothing is changed.

(14.7) The unit translation along the generators of the cylinder $\overline{E}^{n-1} \times \Lambda$ is topological and so is F. Hence $M' \to N'$ defines a topological mapping f of E^{n-1} into a subcell $E_1'^{n-1}$ of E_1^{n-1}. This holds only by virtue of the fact that MN is an arc. Suppose that MN is not an arc, a circumstance which may well arise when γ is closed: N is then some point reached beyond M and MN is the path described along γ. The time of description still being t_1 one may choose an integer k such that if M_i is reached at time it_1/k, $i \leq k$, then the paths M_iM_{i+1} are all arcs. Here of course $M_k = N$. Choose now E_i^{n-1} transverse to γ at M_i. Then if M' is sufficiently near M on E^{n-1} the characteristic γ' through M' will meet E_i^{n-1} in M_i', with $M_k' = N'$ on E_k^{n-1}. By the earlier argument for E^{n-1} sufficiently small $M' \to M_1'$

defines a topological mapping f_1 of E^{n-1} into a subcell $E_1'^{n-1}$ of E_1^{n-1}. Similarly for E_1^{n-1}, hence again for E^{n-1}, sufficiently small $M_1' \to M_2'$ defines a topological mapping f_2 of $E_1'^{n-1}$ into a subcell $E_2'^{n-1}$ of E_2^{n-1}, etc. We thus obtain for E^{n-1} small enough obvious topological mappings f_i and $f = f_k \ldots f_1$ will be a topological mapping of E^{n-1} into a subcell of E_1^{n-1}. This is the result which we had in view.

It has just been assumed that N follows M on γ. By changing t into -t the same results would be obtained when N precedes γ. We may therefore state:

(14.8) Let N follow [precede] M on the characteristic γ and let E^{n-1}, E_1^{n-1} be transverse to γ at M, N. There can be chosen a subcell E'^{n-1} of E^{n-1} such that if $M' \in E'^{n-1}$ then the characteristic γ' through M' followed forward [backward] from M' will first meet E_1^{n-1} in a point N' and $M' \to N'$ defines a topological mapping f of E'^{n-1} into a subcell of E_1^{n-1}

CHAPTER III

LINEAR SYSTEMS

§1. VARIOUS TYPES OF LINEAR SYSTEMS

1. By linear systems are meant those reducible to the form

$$\frac{d\vec{x}}{dt} = A(t)\vec{x} + \vec{b}(t). \tag{1.1}$$

They include the highly important type

$$\frac{d^n x}{dt^n} + a_1(t)\frac{d^{n-1}x}{dt^{n-1}} + \dots + a_{n+1}(t) = 0. \tag{1.2}$$

The special systems (1.1) with $\vec{b} = 0$, or (1.2) with $a_{n+1} = 0$ are known as homogeneous. That is to say they are the systems

$$\frac{d\vec{x}}{dt} = A(t)\vec{x} \tag{1.3}$$

$$\frac{d^n x}{dt^n} + a_1(t)\frac{d^{n-1}x}{dt^{n-1}} + \dots + a_n(t)x = 0. \tag{1.4}$$

We shall deal primarily with real systems (1.3), (1.4).

(1.5) Variation equations. Consider the general system

$$\frac{d\vec{x}}{dt} = \vec{p}(\vec{x},t) \tag{1.5.1}$$

or with components separated:

$$\frac{dx_i}{dt} = p_i(x_1,\dots,x_n,t). \tag{1.5.2}$$

Suppose that we have obtained a solution

$\vec{x}(t) = \{x_i(t)\}$ representing a trajectory over $I \times U$; $I : t_1 < t < t_2$; $U : |x_i - x_i^0| < \alpha$, and that it may be extended to one over $I_1 \times U_1$ where $\bar{I} \subset I_1$, $\bar{U} \subset U_1$. Consider now possible trajectories very near the given one, in the sense that any one of them is of the form $\vec{x}(t) + \vec{\xi}(t)$, and over $I_1 \times U_1$, and that the norm $\| \vec{\xi}(t) \|$ is so small that the products of the ξ_i may be neglected. If we assume that the partial derivatives $\partial p_i / \partial x_j$ exist and are continuous for $(t, \vec{x}(t)) \in I \times U$, then we have from (1.5.2), after neglecting the products of the ξ_i:

$$\frac{d\xi_i}{dt} = \sum \frac{\partial p_i}{\partial x_j} \xi_j . \tag{1.5.3}$$

The system (1.5.3) is known as the system of variation equations of (1.5.1), or of (1.5.2). Similarly if we have

$$\frac{d^n x}{dt^n} = F(x, \frac{dx}{dt}, \ldots, \frac{dx^{n-1}}{dt^{n-1}}, t) \tag{1.5.4}$$

and $x^0(t)$ is a solution, then the consideration of the "neighboring" solutions $x^0(t) + \xi(t)$, where the squares and products of ξ and its derivatives are neglected, leads to a variation equation, obtained thus: if we set

$$\left(\frac{d^k x}{dt^k}\right)_{x=x^0} = y_{k+1}, \qquad \left(\frac{d^0 x}{dt^0}\right)_{x=x^0} = x^0 = y_1 \ :$$

then the equation is

$$\frac{d^n \xi}{dt^n} - \sum \frac{\partial F}{\partial y_k} \frac{d^{k-1}\xi}{dt^{k-1}} = 0 \tag{1.5.5}$$

where $\dfrac{d^0 \xi}{dt^0} = \xi$.

The common feature of the variation systems is that

they are *linear homogeneous*. Thus frequently even if the original system is beyond our scope, once a particular solution is know, the passage to the variation equations may serve to obtain important information regarding the relation of the solution in question to the neighboring solutions.

§2. HOMOGENEOUS SYSTEMS

2. Consider the system (1.3) real, and let R be the interior of the set of all points t where A(t) is continuous. Since the $\frac{\partial p_i}{\partial x_j}$ of (II, 5.2) are here the $a_{ij}(t)$ and the components of R are intervals I, the domain of (1.3) is of the form $I \times \mathcal{V}$. Hence by (II, 10.2):

(2.1) *If* $(\vec{x}^0, t^0) \in I \times \mathcal{V}$ *there passes a unique trajectory through* $(\vec{x}^0, t^0)$. *If* A(t) *is continuous for* $t > \tau$ *there passes a unique trajectory through* $(\vec{x}^0, t^0)$ *whatever* $t^0 > \tau$ *and whatever* $x^0 \in \mathcal{V}$.

(2.2) *If* A *is analytic at* t^0 *then the solution* $\vec{x}(\vec{x}', t)$ *considered as a function of the initial vector* $\vec{x}'$ *is analytic at* $(\vec{x}^0, t^0)$ (see II, 7.2).

(2.3) *If* A *is a function of the vector* $\vec{y}$ *and is analytic at* $(\vec{y}^0, t^0)$ *then the solution is analytic in* $(\vec{x}', t, \vec{y})$ at $(\vec{x}^0, t^0, \vec{y}^0)$. (See II, 7.2).

(2.4) *If* t_1, t_2 *are the end points of* I, *then the boundary of the domain* $I \times \mathcal{V}$ *of* (1.3) *is* $t_1 \times \mathcal{V} \cup t_2 \times \mathcal{V}$. *Therefore* (II, 7.2) *if* $t^0 \in I$, *whatever* $\vec{x}^0$, *any trajectory reaching* $(\vec{x}^0, t^0)$ *can be extended beyond*.

(2.5) *Every solution of* (1.3) *may be extended over the whole interval* I.

For it $\vec{x}(t)$ is a solution its extension can only end at a boundary point of its domain $\mathcal{D}$. By (2.4) such a point is of the form $(\vec{x}, t^1)$, or $(\vec{x}, t^2)$ where t^1, t^2 are the end points of I. Hence the extension cannot be stopped at any point $(\vec{x}, t^0)$, $t^0 \in I$ and this implies (2.5).

3. (3.1) <u>The solutions of</u> (1.3) <u>make up an</u> n-<u>dimensional vector space</u>.

If $\vec{x}(t)$, $\vec{y}(t)$ are two solutions and k, l scalars then $k\vec{x} + l\vec{y}$ is likewise a solution since

$$\frac{d}{dt}(k\vec{x} + l\vec{y}) = kA\vec{x} + lA\vec{y} = A(k\vec{x} + l\vec{y}).$$

Therefore the solutions make up a vector space $\mathcal{W}$. Let $t^0 \in I$, where the domain of (1.3) is $\mathcal{D} = I \times \mathcal{V}_x$ and let $\vec{x}^1, \ldots, \vec{x}^n$ be n independent points of $\mathcal{V}_x$. By (II, 4.1) there is a solution $\vec{x}^i(t)$ such that $\vec{x}^i(t^0) = \vec{x}^i$. To prove (3.1) it is sufficient to show that $\{\vec{x}^i(t)\}$ is a maximal set of linearly independent solutions, or that:

(3.1.1) <u>the</u> $\vec{x}^i(t)$ <u>are linearly independent;</u>

(3.1.2) <u>they span all the solutions</u>, i.e. <u>every solution is of the form</u> $\sum k_i\vec{x}^i(t)$.

If the $\vec{x}^i(t)$ are linearly dependent there must exist a non-trivial identical relation

$$\sum k_i\vec{x}^i(t) = 0$$

valid for all $t \in I$. Hence for $t = t^0$:

$$\sum k_i\vec{x}^i(t^0) = \sum k_i\vec{x}^i = 0$$

which violates the assumption that the $\vec{x}^i$ are linearly independent. This proves (3.1.1).

If $\vec{x}(t)$ is any solution, by (2.5) $\vec{x}(t^0)$ exists whatever $t^0 \in I$. Since $\vec{x}^0$ is in $\mathcal{V}_x$ and $\{\vec{x}^i\}$ is a base for the space we will have

$$\vec{x}^0 = \sum k_i\vec{x}^i.$$

It follows that $\sum k_i\vec{x}^i(t)$ is a solution of (1.3) taking likewise the value $\vec{x}^0$ for $t = t^0$. Since by the fundamental existence theorem (II, 4.1) there is only one such solution we must have

$$\vec{x}(t) = \sum k_h\vec{x}^h(t) \tag{3.2}$$

and this is (3.1.2). Thus (3.1) is proved.

A maximum linearly independent set of solutions $\{\vec{x}^h(t)\}$ for (1.3) is known as a <u>base</u> for (1.3).

(3.3) <u>A n.a.s.c in order that a set of</u> n <u>solutions</u> $\{\vec{x}^h(t)\}$ <u>be a base is that every solution</u> $\vec{x}(t)$ <u>may be represented uniquely in the form</u> (3.2) <u>where the</u> k_i <u>are scalars</u>.

This is a consequence of the definition of the base and of the fact that dim $\mathcal{W} = n$.

(3.4) <u>A n.a.s.c in order that</u> $\{\vec{x}^h(t)\}$ <u>be a base is that if</u> $x_{ih}(t)$ <u>are the coordinates of</u> $\vec{x}^h(t)$, <u>then for some</u> $t^0 \in I$, <u>and hence for every</u> $t \in I$ <u>the determinant</u>

$$D(t) = |x_{ih}(t)| \neq 0. \tag{3.4.1}$$

Suppose $\{\vec{x}^h(t)\}$ is a base. If $\vec{x}(t)$ is any non-trivial solution we will have a non-trivial relation (3.2) and hence for $t = t^0$:

$$x_i(t^0) = \sum k_h x_{ih}(t^0). \tag{3.5}$$

Thus (3.5) considered as a system of n equations of the first degree in the unknowns k_h, does have a solution. Furthermore the solution is unique. For regardless of $\vec{x}(t)$, if we are given $\vec{x}(t^0) = \vec{x}^0$, then corresponding to any solution $(k_1,\ldots,k_n)$ of (3.5) the function $\sum k_i x^i(t)$ will be a solution of (1.3) taking the value $\vec{x}^0$ for $t = t^0$. Since the function is unique (II, 4.1) so is the set $(k_1,\ldots,k_n)$. It follows that $D(t^0) \neq 0$. Since t^0 is any $t \in I$, (3.4.1) holds and so the condition of (3.4) is necessary.

Suppose now that $D(t^0) \neq 0$ for some $t^0 \in I$. By (2.5) $\vec{x}(t^0)$ is defined. Under our assumption (3.5) will then have a unique solution $(k_1,\ldots,k_n)$ and so $\sum k_h \vec{x}^h(t)$ will be a solution assuming the same value $\vec{x}(t^0)$ for $t = t^0$ as $\vec{x}(t)$. Since such a solution is unique there subsists a relation (3.2). Thus $\{\vec{x}^h(t)\}$ spans all the solutions and since it consists of n vectors it is a base. By the necessity of the condition (3.4.1) will then hold for every $t \in I$. This proves (3.4).

(3.6) The transformation $\vec{x}^0 = \vec{x}(t^0) \to \vec{k}$ resulting from (3.5) is a topological mapping $\mathcal{V}_x \to \mathcal{V}_k$. Hence $\sum k_h \vec{x}^h(t)$ is a general solution.

(3.7) <u>A noteworthy property of the variation equation</u>. Consider an equation (1.5.3) with $\vec{p}$ analytic in $\vec{x}$ and a domain of analyticity $\Delta = L \times \Delta_1$, where L is the real t line and Δ_1 a region of $\mathcal{V}_x$. Suppose also that we have a solution $\vec{x}(t,\vec{c})$, $\vec{c} = (c_1,\ldots,c_r)$, analytic in $\vec{c}$ for $\vec{c}$ in a certain region Λ of $\mathcal{V}_c$, and all t, and with a Jacobian matrix J of rank r in Λ. In particular for a certain $c^0 \in \Lambda$, $\vec{x}(t,\vec{c}^0) = \vec{x}(t)$, the solution of (1.5.2) whose variation equation is (1.5.3). Then:

(3.8) <u>The</u> r <u>functions</u> $\dfrac{\partial \vec{x}(t,\vec{c})}{\partial c_i}$ <u>are linearly independent solutions of the variation equation</u> (1.5.3).

Linear independence is an immediate consequence of the fact that the Jacobian matrix J is of rank r, so we only need to prove that the functions in question are solutions. Let $\Delta\vec{c} = (0,\ldots,0, \Delta c_i, 0,\ldots)$. Since $\vec{x}(t,\vec{c}^0+\Delta\vec{c})$ is a solution whatever Δc_i, if we substitute for the x_h in (1.5.2) the expansions

$$x_h(t,\vec{c}^0+\Delta\vec{c}) = x_h(t,\vec{c}^0) + \frac{\Delta c_i}{1!}\,\frac{\partial x_h(t,\vec{c}^0)}{\partial c_i^0} + \cdots$$

the result must be an identity in Δc_i. If we identify the powers of Δc_i on both sides we find that the vector whose components are $\dfrac{\partial x_h(t,c^0)}{\partial c_i^0}$ is a solution of the variation equation. This proves (3.8).

<u>Remark</u>. The argument could be extended without particular difficulty to the non-analytical case.

4. (4.1) Side by side with (1.3) it is interesting to consider the associated matrix equation

$$(4.2)\qquad \frac{dX}{dt} = AX.$$

Written out explicitly it takes the form

$$(4.3)\qquad \frac{dx_{ih}(t)}{dt} = \sum a_{ij}(t)x_{jh}(t).$$

It shows that the vector $\vec{x}^h$ whose coordinates are the elements $x_{jh}(t)$ of the hth column of X, is a solution of (1.3). Thus the columns of X give rise to n solutions of the differential equation (1.3). Further developments rest upon the important relation

$$(4.4)\qquad | X(t) | = | X(t^0) | \exp (\int_{t^0}^{t} (\mathrm{tr}\, A)dt),$$

for all t^0, $t \in I$. The proof is as follows: By a well known rule for the derivative of a determinant we find

$$\frac{d\ |X(t)|}{dt} = \sum_i \left| \frac{dx_{ik}}{dt} \right| = \sum_i \left| \sum_j a_{ij}x_{jk} \right| = \sum_{i,j} \left| a_{ij}x_{jk} \right|$$

where the terms unwritten in each determinant are as in X itself. In the last determinant there are proportional rows unless $i = j$. Hence terms with $i = j$ are the only terms which do not vanish and they have the values $a_{ii}\ |X|$. Hence

$$\frac{d\ |X|}{dt} = |X|\ \mathrm{tr}\ A$$

from which (4.4) follows by integration.

An immediate consequence of (4.4) is

(4.5) If $|X(t)| \neq 0$ for some $t \in \bar{I}$ then it is $\neq 0$ for all $t \in I$.

A matrix-solution of (4.2) such that $|X(t)| \neq 0$ is called a non-singular solution.

(4.6) If $\vec{x}^h$ is the solution of the differential equation (1.3) represented by the hth column of the matrix solution X of (4.2) then a n.a.s.c in order that $\{\vec{x}^h(t)\}$ be a base for (1.3) is that X be a non-singular solution of (4.2).

This is an immediate consequence of (3.4).

(4.7) Remark. It is clear now that (4.5) is likewise implicit in (3.4). However the basic relation (4.4) is important for its own sake; furthermore there is some interest in having a direct proof of the type given here, rather than one such as for (3.4), based upon the existence theorem.

(4.8) Owing to our habitual identification of vectors with one-column matrices we have naturally associated with (1.3) the matrix equation (4.2) with X as a right multiplier. We could equally identify vectors with one-row matrices and we would then associate in place of (1.3), (4.2) the equations

$$\frac{d\vec{x}}{dt} = \vec{x}B \tag{4.8.1}$$

$$\frac{dX}{dt} = XB \tag{4.8.2}$$

and everything said so far would hold with rows and columns interchanged.

5. Adjoint systems. The special vector and matrix equations

$$\frac{d\vec{y}}{dt} = -\vec{y}A \tag{5.1}$$

$$\frac{dY}{dt} = -YA \tag{5.2}$$

where A is the same as in (1.3) and (4.2) are said to be adjoint to (1.3) and (4.2). Since

$$\frac{dY'}{dt} = (-A')Y', \tag{5.1)'$$

$$\frac{dX'}{dt} = -X'(-A'), \tag{1.3)'$$

(1.3)' may be considered as the adjoint system to (5.1)'.

This brings out the symmetry in the situation. If the matrix A is <u>skew-symmetric</u> then $A = -A'$ ($a_{ii} = 0$, $a_{ij} = -a_{ji}$) and (4.2) is the same as (5.1)', (5.2) the same as (1.3)'. All the systems considered are then said to be <u>self-adjoint</u>.

From (4.2) and (5.2) there follows:

(5.3) $$Y\frac{dX}{dt} + \frac{dY}{dt}X = 0$$

and therefore

(5.4) $$YX = C,$$

where C is a matrix of scalars. Let Y be a solution of (5.2) such that $|Y| \neq 0$ and set $X = Y^{-1}C$. Then (5.4) holds, and hence also (5.3). Consequently

(5.5) $$Y\frac{dX}{dt} = +YAX.$$

Since $|Y| \neq 0$, Y^{-1} exists and so from (5.5) follows (4.3). Consequently every solution of (4.2) is of the form $Y^{-1}C$. This proves:

(5.6) <u>If Y is a non-singular solution of the adjoint equation to (4.3) then every solution X of (4.3) is represented by $Y^{-1}C$ where C is an arbitrary scalar matrix. The non-singular solutions correspond to $|C| \neq 0$.</u>

Taking $C = E$ we have $YX = E$, and so X^{-1} is a special solution of (5.2). Since $Y^{-1} = X$, we recognize in (5.6) proposition (3.3) in another formulation.

(5.7) For obvious reasons of symmetry the same holds with (4.3) replaced by (5.2), Y by X and $Y^{-1}C$ by CX^{-1}.

(5.8) In the applications the important associated types are really (1.3) and (4.2). For this reason it will be worth while to be a little more explicit regarding the relation of a system (1.3) to its adjoint. If we write (1.3) as

(5.8.1) $$\frac{dx_i}{dt} = \sum a_{ij}x_j,$$

then the adjoint (5.1) assumes the form

$$\frac{dy_i}{dt} = -\sum a_{ij}y_i. \tag{5.8.2}$$

If $(x_1,\ldots,x_n)$, $(y_1,\ldots,y_n)$ are any two solutions of the respective systems then clearly

$$\sum (x_i dy_i + y_i dx_i) = 0 \tag{5.8.3}$$

and so

$$\sum x_i y_i = C, \tag{5.8.4}$$

a constant. This is the analogue of (5.4), and could in fact be deduced from it. If we have a base $\{\vec{y}^j\}$, $\vec{y}^j = (y_{j1},\ldots,y_{jn})$, for the solutions of the adjoint to (1.3) then $|y_{ji}| \neq 0$. From (5.8.4) follows also

$$\sum y_{ji}x_i = C_j, \tag{5.8.5}$$

a system of equations of the first degree which may be solved for the x_i. The solution thus obtained is in terms of n arbitrary constants, the C_j, and is the general solution of (5.8.1). We may thus state:

(5.9) Given a linear homogeneous system of differential equations (5.8.1) if we have n linearly independent solutions of the adjoint system (5.8.2), then the complete solution of the system (5.8.1) itself is reduced to the solution of an algebraic system of n linear equations in n unknowns.

6. The linear homogeneous differential equation of order n. Since (1.4) is reducible to a system (1.3) its properties may be deduced from those of (1.3). For convenience in the applications we shall express them directly.

For the sake of expediency we introduce the customary operator $D = \frac{d}{dt}$. Then (1.4) takes the form

$$(6.1)\qquad D^n x + a_1(t)D^{n-1}x + \dots + a_n(t)x = 0,$$

or with the conventions $D^0 = 1$, $a_0 = 1$:

$$(6.2)\qquad \sum a_k(t)D^{n-k}x = 0.$$

Writing $x = x_1$, the equivalent system (5.8.1) is:

$$(6.3)\qquad \begin{aligned} Dx_1 &= x_2 \\ &\cdots\cdots \\ Dx_{n-1} &= x_n \\ Dx_n &= -a_n x_1 - a_{n-1}x_2 - \dots - a_1 x_n. \end{aligned}$$

This yields in particular:

$$(6.4)\qquad x_k = D^{k-1}x.$$

The domain of (6.1) or of (6.3) is of the form $I \times \mathcal{V}$, where I is a component of the interior of the set of points of the real t line at which all the $a_i(t)$ are continuous.

Let us look at the question of linear dependence of the solutions of (6.1) at first directly, i.e. without referring to (6.3). A set of solutions $\{\xi^i(t)\}$ is said to be <u>linearly dependent</u> whenever there can be found real scalars C_i not all zero such that

$$(6.5)\qquad \sum C_i \xi^i(t) = 0.$$

When this holds we also have

$$(6.6)_k\qquad \sum C_i D^k \xi^i(t) = 0,$$

for all k. Of course since by (6.1) the $D^n\xi^i$ are linearly dependent upon $\xi^i, \dots, D^{n-1}\xi^i$, (6.6) need only be considered for $k < n$. It is clear that the $\xi^i(t)$ will be linearly dependent whenever $(6.6)_0, \dots, (6.6)_{n-1}$ have a solution in the C_i not all zero, and hence certainly if matrix

$$(6.7)\qquad \| D^k\xi^i \|, \quad (k = 0,\dots,n-1;\ i = 1,2,\dots,r)$$

has more than n columns. Hence there cannot be more than n linearly independent solutions. Given on the other hand n solutions a n.a.s.c. for their linear dependence is that the determinant called <u>Wronskian</u>:

$$\Delta(\xi^1,\dots,\xi^n) = \begin{vmatrix} \xi^1 & \dots & \xi^n \\ D\xi^1 & \dots & \\ \vdots & & \\ D^{n-1}\xi^1 & \dots & \end{vmatrix} = 0.$$

We have already shown (3.3) that there are n linearly independent solutions and no more. A set of n linearly independent solutions ξ^1, ..., ξ^n is known as a <u>base</u> for the solutions. Any other ξ is given by a relation

$$\xi(t) = \sum c_i\xi^i(t).$$

If we think of $\xi(t)$ as a function $\xi(t;c_1,\dots,c_n)$ then it is described as a <u>general</u> solution of (6.1).

Coupling what precedes with the properties of the Wronskian we have:

(6.8) <u>A n.a.s.c. in order that</u> $\{\xi^1,\dots,\xi^n\}$ <u>be a base for the solutions of</u> (6.1) <u>is that the Wronskian</u> $\Delta(\xi^1,\dots,\xi^n) \neq 0$ <u>for some</u> $t^0 \in I$, <u>and hence for every</u> $t \in I$.

(6.9) <u>Given</u> n <u>linearly independent functions</u> ξ^1, ..., ξ^n <u>each</u> n <u>time differentiable on an interval</u> I: $t_1 < t < t_2$, <u>they satisfy the differential equation</u>

$$(6.9.1)\qquad \Delta(x,\xi^1,\dots,\xi^n) = 0$$

<u>and this equation is unique to within a factor in</u> t.

If we replace x by ξ^i, Δ acquires two identical rows and so vanishes. Hence ξ^i satisfies (6.9.1). Unicity will follow if we can prove more generally:

(6.10) <u>If the two equations</u>

$$(6.10.1)\quad D^n x + a_1(t)D^{n-1}x + \dots + a_n^{(t)}x = 0\,.$$

$$(6.10.2)\quad D^n x + b_1(t)D^{n-1} + \dots + b_n(t)x = 0$$

<u>are satisfied over the interval I by the same set of n linearly independent functions then they are identical</u>.

For otherwise the difference is of order $< n$, not identically zero, yet with n linearly independent solutions which is ruled out. Therefore (6.10) holds and so does (6.9).

$$(6.11)\qquad \Delta(t) = \Delta(t^0)\exp\left(-\int_{t_0}^{t} a_1(t)dt\right).$$

In fact by differentiation

$$\frac{d\Delta}{dt} = \begin{vmatrix} \xi^i \\ D\xi^i \\ \vdots \\ D^{n-2}\xi^i \\ D^n\xi^i \end{vmatrix} = - \begin{vmatrix} \xi^i \\ \vdots \\ D^{n-2}\xi^i \\ \sum a_k D^{n-k}\xi^i \end{vmatrix} = -a_1\Delta,$$

from which to (6.11) is but a step.

(6.12) It is not difficult to see that in what precedes the Wronskian matrix plans the role of the previous matrix X. In fact referring to the system (6.3) associated with, and equivalent to (6.1), if we set $D^{k-1}\xi^i(t) = x_{ki}(t)$, then

$$\Delta(\xi^1,\dots,\xi^n) = \| x_{ki}(t) \| = X.$$

To $\xi^i(t)$ there corresponds now the vector solution $\vec{x}^i(t)$ of (6.3) and clearly the correspondence $\xi^i(t) \leftrightarrow x^i(t)$ is one to one and preserves linear dependence. As a consequence the properties just obtained for (6.1) could be obtained directly from those of (6.3).

7. We will now discuss adjoints for the type (6.1).

First the adjoint of (6.3) is

$$
(7.1)\qquad
\begin{aligned}
Dy_1 &= a_n y_n\\
Dy_2 &= -y_1 + a_{n-1} y_n\\
&\cdots\cdots\cdots\\
Dy_n &= -y_{n-1} + a_1 y_n
\end{aligned}
$$

If we differentiate the (k+1)st relation k times and set $y_n = y$ we find

$$(7.2)\qquad D^n y - D^{n-1}(a_1 y) + \dots + (-1)^n a_n y = 0,$$

and we now define (6.1), (7.2) as <u>adjoint</u> to one another. To derive the analogue of (5.8.4) we must solve (7.1) for the y_k. We first have

$$D^i y_{k+1} = -D^{i-1} y_{k+i-1} + D^{i-1}(a_{n-k-i+1} y)$$

and this yields

$$y_k = \sum_{h=0}^{n-k} (-1)^h D^h (a_{n-k-h} y).$$

Substituting finally in (5.8.4) we obtain as the analogue of (5.8.4):

$$(7.3)\qquad \sum_{h,k} (-1)^h D^{k-1} x D^h (a_{n-k-h} y) = C.$$

If we possess a base $\{\eta_i\}$ for the solution of the adjoint equation (7.1) to (6.1) then the <u>general</u> solution x(t) of (6.1) may be obtained by solving the linear (algebraic) system

$$(7.4)\qquad \sum_{h,k} D^{k-1} x D^h (a_{n-k-h} \eta^i) = C_i \qquad (i = 1,2,\dots,n)$$

for x.

(7.5) The relation just obtained for adjoints may also be derived directly by means of integration by parts. For any two functions x(t), y(t) with a suitable number of derivatives we have:

$$y\,D^k x = D(yD^{k-1}x) - DyD^{k-1}x$$

$$\cdots\cdots\cdots\cdots$$

$$D^h y D^{k-h} x = D(D^h y D^{k-h-1}x) - D^{h+1}yD^{k-h-1}x$$

$$\cdots\cdots\cdots\cdots$$

$$D^{k-1}y\,.\,x = D(D^{k-1}yDx) - D^k y\,.\,x.$$

From this follows

$$yD^k x + (-1)^k x D^k y = D\sum(-1)^h D^h y D^{k-h-1}x.$$

Replacing k by n-k and y by $a_k y$ we find:

$$\begin{aligned} y\,.\,a_k D^{n-k}x &+ (-1)^{n-k} x D^{n-k} a_k y \\ &= D\sum(-1)^h D^h(a_k y)D^{n-k-h-1}x. \end{aligned} \tag{7.6}$$

Hence if x is a solution of (6.1) and y a solution of its adjoint,

$$D\sum_{h,k}(-1)^h D^h(a_k y)D^{n-k-h-1}x = 0, \tag{7.7}$$

from which with a rearrangement of indices it is but a step to (7.3).

(7.8) Green's formula. Generally speaking let L(x), M(x) be two linear differential opeators of same order

$$L(x) = \sum a_k(t)D^{n-k}x$$
$$M(x) = \sum b_k(t)D^{n-k}x.$$

We shall say that L,M are adjoint to one another if

$$yL(x) - xM(y) = \frac{dF}{dt} \tag{7.9}$$

where F is a polynomial in x,y and their derivatives of orders $\leq$ n-1 with coefficients functions of t. It is assumed all the way through that all the functions of t under consideration are continuous together with their derivatives as far as required on some interval I. If t^1, $t^2 \in I$ we will then have as a consequence of (7.9):

(7.10) $$\int_{t_1}^{t_2} (yL(x) - xM(y))dt = F \Big]_{t^1}^{t^2},$$

a result known generally as <u>Green's formula</u>. It is particularly useful wherever one deals with preassigned "boundary" conditions at t^1, t^2.

(7.11) <u>Example</u>. If we have

$$L(x) = M(x) = \frac{d}{dt}\left(p \frac{dx}{dt}\right) + qx,$$

then

$$yL(x) - xL(y) = \frac{d}{dt}\left(p\left(y \frac{dx}{dt} - x \frac{dy}{dt}\right)\right)$$

and so L is self-adjoint. We have then

$$\int_{t^1}^{t^2} (yL(x) - xL(y))dt = \left[p\left(y \frac{dx}{dt} - x \frac{dy}{dt}\right)\right]_{t_1}^{t_2}.$$

Thus the integral at the left depends only on the values of the functions and their derivatives at the end points t^1, t^2.

§3. NON-HOMOGENEOUS SYSTEMS

8. Consider first the type

(8.1) $$\frac{d\vec{x}}{dt} = A(t)\vec{x} + \vec{b}(t).$$

The adjoint of the associated homogeneous equation is

(8.2) $$\frac{d\vec{y}}{dt} = -\vec{y}\, A(t).$$

Let $\{\vec{y}^i\}$ be a base for the solutions of (8.2) with $\vec{y}^i = \{y_{ij}\}$, $Y = \| y_{ij} \|$, and so $|Y| \neq 0$. In particular Y^{-1} will exist. Since

(8.3) $$\frac{dY}{dt} = -YA$$

we will have

$$Y \frac{d\vec{x}}{dt} + \frac{dY}{dt} \vec{x} = Y\vec{b} = \frac{d(Y\vec{x})}{dt}$$

Hence

(8.4) $$Y\vec{x} = \int_{t_0}^{t} Y\vec{b}\, dt, \quad \vec{x} = Y^{-1} \int_{t_0}^{t} Y\vec{b}\, dt.$$

Similarly the matrix differential equation

(8.5) $$\frac{dX}{dt} = AX + B$$

has the solution

(8.6) $$X = Y^{-1} \int_{t_0}^{t} YB\, dt.$$

The same method may be applied to the solution of the non-homogeneous equation

(8.7) $$\sum a_k D^{n-k} x = b(t).$$

If y satisfies the adjoint equation (7.2) to the related homogeneous equation then (7.6) yields this time

(8.8) $$by = \sum_{h,k} (-1)^h D^h (a_k y) D^{n-k-h-1} x.$$

If we possess a base $\{\eta^i\}$ for the solutions of (7.2), the related relations (8.8) duly integrated yield

(8.9) $$\int_{t_0}^{t} b(t)\eta^i(t)dt = \sum (-1)^h D^h (a_k \eta^i) D^{n-k-h-1} x.$$

This is a linear (algebraic) system in x, Dx, ..., $D^{n-1}x$, whose determinant is of the form $f(t)\, W(\eta^1, \ldots, \eta^n)$, $f(t) \neq 0$, and so it may be solved for x.

(8.10) In each of the three types just considered we have succeeded in obtaining one solution. The general solution is obtained by adding the general solution of the corresponding homogeneous system. Take for instance (8.7). If x, x' are two solutions then $z = x' - x$ satisfies the homogeneous equation

(8.11) $$\sum a_k D^{n-k} z = 0.$$

Hence $x' = x + z$, where z is the general solution of (8.11), and x' is the general solution of (8.7).

9. Application. Consider the system

(9.1) $$\frac{d^2x}{dt^2} + x = b(t).$$

It is simpler here to operate directly with the equivalent system of two equations of the first order:

$$\frac{dx}{dt} = y$$

$$\frac{dy}{dt} = -x + b(t).$$

We have here

$$Y = Y^{-1} = \begin{Vmatrix} \sin t & \cos t \\ \cos t & -\sin t \end{Vmatrix}$$

and therefore by (8.4)

$$(x,y) = (x, \frac{dx}{dt}) = Y \int_0^t Y(0,b)dt$$

$$= \begin{Vmatrix} \sin t & \cos t \\ \cos t & -\sin t \end{Vmatrix} \int_0^t \begin{Vmatrix} \sin u & \cos u \\ \cos u & -\sin u \end{Vmatrix} (0,b(u))du$$

$$= \begin{Vmatrix} \sin t & \cos t \\ \cos t & -\sin t \end{Vmatrix} \int_0^t (b \cos u, - b \sin u)\, du$$

$$= \int_0^t (b \sin(t - u), b \cos (t-u))\, du$$

and therefore explicitly

(9.2) $$x = \int_0^t b(u) \sin(t-u)du, \quad \frac{dx}{dt} = \int_0^t b(u) \cos(t-u)du.$$

This is the well known form of the solution of (9.1) which vanishes, together with its derivative, for $t = 0$.

Of course the expression of $\frac{dx}{dt}$ may be obtained from that of x by differentiation under the integration sign. The general solution is obtained by adding to x the complementary function, and is explicitly given by:

$$(9.3) \qquad x = C \cos (t-\alpha) + \int_0^t b(u) \sin(t-u)du$$

where C, α are arbitrary constants.

§4. LINEAR SYSTEMS WITH CONSTANT COEFFICIENTS

10. Suppose that in

$$(10.1) \qquad \frac{d\vec{x}}{dt} = A\vec{x},$$

A is a constant. The associated matrix equation

$$(10.2) \qquad \frac{dX}{dt} = AX$$

has already been solved (I, 7.7) and found to have the solution

$$(10.3) \qquad X = e^{At}$$

which is non-singular (I, 6.2). By (4.6) if $\vec{x}^h = (x_{1h}, \ldots, x_{nh})$, the set $\{\vec{x}^h\}$ is a base for (10.1), and the solution is thus complete.

It is of interest to examine the form of the solution. We first observe that if we apply the transformation $\vec{x} = P\vec{y}$, $|P| \neq 0$, then (10.1) is replaced by

$$(10.4) \qquad \frac{d\vec{y}}{dt} = B\vec{y}, \qquad B = P^{-1}AP.$$

We choose P so that B is canonical. In place of (10.2) we have now

$$(10.5) \qquad \frac{dY}{dt} = BY,$$

with the non-singular solution e^{Bt}. Notice now that

$e^{At} = Pe^{Bt}P^{-1}$, (I, 5.10), hence we merely need to find the form of e^{Bt}. Now if $B = \mathrm{diag}(B_1, \ldots, B_r)$ then $e^{Bt} = \mathrm{diag}(e^{B_1 t}, \ldots, e^{B_r t})$, (I, 5.4). If

$$(10.6) \qquad B_h = \left\| \begin{matrix} \lambda_j & & & 0 \\ 1 & \lambda_j & & \\ & \ddots & \ddots & \\ 0 & & 1 & \lambda_j \end{matrix} \right\|$$

the same calculation as in (I, 5.6) will yield if ρ_h is the order of B_h:

$$(10.7) \qquad e^{B_h t} = e^{\lambda_j t} \left\| \begin{matrix} 1 & & & & 0 \\ \frac{t}{1!} & 1 & & & \\ \frac{t^2}{2!} & \frac{t}{1!} & 1 & & \\ \cdots & \cdots & \cdots & \ddots & \\ \frac{t^{\rho_h - 1}}{(\rho_h - 1)!} & \cdots & \cdots & \frac{t}{1!} & 1 \end{matrix} \right\|$$

and e^{Bt} is made up of these diagonal blocks.

(10.8) Special case: $A \sim \mathrm{diag}(\lambda_1, \ldots, \lambda_n)$. Then clearly $Y = \mathrm{diag}(e^{\lambda_1 t}, \ldots, e^{\lambda_n t})$ and so it has a base $\{\vec{y}^h\}$ where

$$(10.9) \qquad \vec{y}^h = (\delta_{hj} e^{\lambda_j t}) = (0, \ldots, 0, e^{\lambda_h t}, 0, \ldots, 0),$$

so that $\vec{y}^h$ has a single coordinate $\neq 0$ in the hth place. The general solution of (10.4) is

$$(10.10) \qquad \vec{y} = (C_1 e^{\lambda_1 t}, C_2 e^{\lambda_2 t}, \ldots, C_n e^{\lambda_n t}).$$

The general solution for $\vec{x}$ is $\vec{x} = P\vec{y}$ or:

$$x_i = \sum p_{ij} C_j e^{\lambda_j t} \qquad (i = 1,2,\ldots,n).$$

(10.11) In point of fact it is not particularly difficult to describe the solutions in the general case. We will begin with the $\vec{y}^i$. Set $\sigma_h = \rho_1 + \cdots + \rho_{h-1}$. Let also y_{mk} denote as usual the coordinates of $\vec{y}^k$. Then we have

$$y_{mk} = \frac{t^{m-k}}{(m-k)!} e^{\lambda_j t}, \ \sigma_h < k \leq m \leq \sigma_{h+1},$$

and all other coordinates y_{mk} are zero.
Hence $\vec{x}^k$ will have its coordinates x_{ik} of the form

$$x_{ik} = \varphi_{ik}(t) e^{\lambda_j t}$$

where φ_{ik} is a polynomial in t of degree $\leq \rho_{h-1}$. The general solution $\vec{x}$ will then have its coordinates x_i given by

$$x_i = \sum C_k \varphi_{ik}(t) e^{\lambda_j t}$$

where the C_k are arbitrary constants. If B_h is of order one the φ_{ik}, with $\sigma_h < k \leq \sigma_{h+1}$, are all equal to unity, and the corresponding terms in the general solutions are $C_k e^{\lambda_j t}$

(10.12) Let us suppose explicitly that the characteristic roots are $\lambda_1, \bar{\lambda}_1, \ldots, \lambda_r, \bar{\lambda}_r, \lambda_{2r+1}, \ldots, \lambda_n$ where the λ_{2r+i} are real. Then (10.1) may be reduced to the form (10.4) with the λ's forming a system $(\lambda_1, \bar{\lambda}_1)$, $\ldots, (\lambda_r, \bar{\lambda}_r), \lambda_{2r+1}, \ldots, \lambda_n$ with r conjugate pairs and the rest real. The coordinates may likewise be chosen such that for real points they form sets $(x_1, \bar{x}_1, \ldots, x_r, \bar{x}_r, x_{2r+1}, \ldots, x_n)$ with the last n-2r real.

The modifications in (10.11) are quite simple. We merely note that the general real solution when the roots are distinct is

$$(10.12.1)\quad x_k = \sum_{i=1}^{r} (c_i e^{\lambda_i t} + \bar{c}_i e^{\bar{\lambda}_i t}) + \sum_{i=2r+1}^{n} c_i e^{\lambda_i t}$$

when the c_i, $i > 2r$, are real.

§5. LINEAR SYSTEMS WITH PERIODIC COEFFICIENTS

11. Consider again our basic system (1.3), or

$$(11.1)\qquad \frac{d\vec{x}}{dt} = A\vec{x}.$$

We suppose t real and A real, continuous for all t, and with the real non-zero period T. Thus $A(t+T) = A(t)$. We will allow however for convenience <u>complex</u> solutions of (11.1), and so complex scalars. If L is the real line and $\mathcal{V}$ a complex n dimensional vector space the domain of (11.1) is $L \times \mathcal{V}$.

Let $X = \| x_{ij} \|$, $X \neq 0$ be a non singular solution of

$$(11.2)\qquad DX = \frac{dX}{dt} = AX.$$

Thus if $\vec{x}^j = (x_{1j}, \ldots, x_{nj})$, then $\{\vec{x}^j\}$ is a base for the solutions of (11.1). Evidently $X(t+h)$ satisfies (11.2) for the value $t + T$ of t and as we know $|X(t+T)| \neq 0$ also. Now

$$DX(t+T) = A(t+T)\, X(t+T) = A(t)\, X(t+T).$$

Hence $X(t+T)$ is a non-singular solution of (11.2) with non zero-determinant, for the value t itself. From this follows

$$(11.3)\qquad X(t+T) = X(t) \cdot C$$

where C is constant. Since $|X(t)| \neq 0$, $X^{-1}(t)$ exists, $C = X^{-1}(t)\, X(t+T)$ and $|C| \neq 0$. In fact by (4.6):

$$(11.3.1) \qquad |C| = e^{\int_0^T (\text{tr. } A)\, dt}$$

(11.4) If we replace $\{\vec{x}^h\}$ by another base the effect is to replace X by XP, $|P| \neq 0$, and hence C by $P^{-1}X^{-1}(t)\, X(t+T)P = P^{-1}CP \sim C$. Since P is an arbitrary non-singular matrix and the choice of base is essentially immaterial, we may assume X such that C is a canonical matrix: $C = \text{diag}\,(C_1, \ldots, C_r)$, where C_i is of the form (I, 3.6.1). In particular if the characteristic roots $\lambda_1, \ldots, \lambda_n$ of C are all distinct we will have $C = \text{diag}\,(\lambda_1, \ldots, \lambda_n)$.

(11.5) Since $|C| \neq 0$, likewise $|C_i| \neq 0$. Hence (I, 6) there exists a matrix B_i of the same order as C_i such that $e^{TB_i} = C_i$ and if $B = \text{diag}\,(B_1, \ldots, B_r)$, then $e^{TB} = C$.

(11.6) Notice that if the characteristic roots are all distinct, or if they are not and still $C = \text{diag}\,(\lambda_1, \ldots, \lambda_n)$ then choosing for μ_i any determination of $\frac{1}{T} \log \lambda_i$, we may take $B = \text{diag}\,(\mu_1, \ldots, \mu_n)$. If the λ_i are all distinct the same will be true as regards the μ_i.

(11.7) Having chosen B consider the matrix

$$(11.8) \qquad Z(t) = e^{tB} \,.\, X^{-1}.$$

If we recall that B and its power series commute we find

$$\begin{aligned} Z(t+T) &= e^{(t+T)B} \,.\, (X^{-1}(t+T)) = e^{TB} \,.\, e^{tB}(X(t)C)^{-1} \\ &= e^{tB} \,.\, e^{TB} \,.\, C^{-1} \,.\, X^{-1}(t) \\ &= e^{tB}\, X^{-1}(t) = Z(t). \end{aligned}$$

In other words Z(t) has the period T. Moreover by (I, 6.2):

$$(11.9) \qquad |Z(t)| = |e^{tB}| \cdot |X^{-1}(t)| \neq 0$$

for all values of t. Setting $\overline{X} = ZX = e^{tB}$ (here and below $\overline{X}$, ..., is not the conjugate of X, ...) then

$$(11.10) \qquad \frac{d\overline{X}}{dt} = B\overline{X}.$$

The relations between the coordinates assume the form:

$$\overline{x}_{ih} = \sum z_{ij}(t)x_{jh}$$

Here $\overline{x}_{ih}$ are the coordinates of the vector $\vec{x}^h$ referred to the coordinate system $(\overline{x}_1, \ldots, \overline{x}_n)$ deduced from the initial system by the linear transformation

$$(11.11) \qquad \overline{x}_i = \sum z_{ij}(t)\, x_j,$$

which is non-singular owing to (11.9). The relation (11.10) yields

$$(11.12) \qquad \frac{d\overline{x}_{ih}}{dt} = \sum b_{ij}\overline{x}_{jh}.$$

Thus $\vec{\overline{x}}^h = (\overline{x}_{1h}, \ldots, \overline{x}_{nh})$ is a solution of the linear homogeneous system with constant coefficients

$$(11.13) \qquad \frac{d\overline{x}_i}{dt} = \sum b_{ij}\overline{x}_j$$

(11.14) Since $|\overline{X}| = |\overline{x}_{ij}| = |Z| \cdot |X| \neq 0$, the set $\{\vec{\overline{x}}^h\}$ is a base for (11.3). The elements of this base may be written

$$\overline{x}_{ih} = \sum \varphi_{jh}(t)\, e^{\mu_j t}$$

where the μ_j are the characteristic roots of B, and the φ's are polynomials. Whenever the μ_j are all distinct then the φ's are constant. It follows that the initial base $\{\vec{x}^h(t)\}$ for (11.1) is of the form

$$x_{ih} = \sum \psi_{jh}(t)\, e^{\lambda_j t}$$

where the ψ's are polynomials in t with coefficients periodic and of period T, or else if the λ_j are all distinct merely periodic functions of T. The λ_j are known as the characteristic exponents of the differential equation (11.1). Thus

(11.15) Theorem. By a transformation of variables (11.11) the periodic system (11.1) may be reduced to a linear homogeneous system with constant coefficients.

The characteristic roots of B are the numbers μ_i, and referring to (10.6) they are distinct when the λ_i themselves are distinct. By (9) and (11.15) we have:

(11.16) There is a base $\{\vec{x}^h\}$ for the solutions of the periodic system (11.1) whose n elements are of the form

$$\vec{x}^h = \{\xi_{1h}(t)e^{\lambda_1 t}, \ldots, \xi_{nh}(t)e^{\lambda_n t}\}.$$

(11.17) When the characteristic exponents λ_i are all distinct, the solutions are all of the form $\sum \psi_j(t)e^{\lambda_j t}$ where ψ_i is periodic and of period T.

(11.18) Real solutions. All that is required is to obtain a real base, and a process for the purpose has been developed in (9.3).

CHAPTER IV

CRITICAL POINTS AND PERIODIC MOTIONS

QUESTIONS OF STABILITY

By critical point we shall mean a point $\vec{x}^0$ such that $(\vec{x}^0, t)$ is a solution. In the present chapter we shall discuss the representations of the solutions in the vicinity of critical points and of periodic motions and take up also the important related questions of stability. These problems were investigated above all by Poincaré and Liapounoff. Our treatment rests primarily upon the classical (but not well known) Mémoire of Liapounoff (original dated 1892 in Russian; translated under the title: Probleme general de la stabilite du movement, Annales de Toulouse, ser 2, vol. 9 (1907), pp. 203-475). The chief topic of the paper is the investigation of stability of certain trajectories, but it contains a number of noteworthy results along different directions. Poincaré's results are found in his Thèse, and in his treatise: Méthodes nouvelles de la Mecanique Celeste, vols. I and III. Some of them are dealt with in Picard's Traite d'Analyse, vol. III, and there are several noteworthy papers due to Horn (Journal fur Math., vol. 116 (1896) and subsequent volumes) on the same questions.

§1. STABILITY

1. This concept dominates the work of Liapounoff and motivates more than anything else the by no means simple analysis to follow.

(1.1) Historically the question of stability was first raised by Lagrange in connection with the equilibrium of conservative systems. For our present purpose it is sufficient to recall that the state of a conservative system $\sum$ depends upon a certain real vector $\vec{x}$ and its first derivative $\vec{x}'$ and more particularly upon two continuous real functions:

the kinetic energy $T(\vec{x},\vec{x}')$ which is positive save that $T = 0$ when and only when $\vec{x}' = 0$,

the potential energy $V(\vec{x})$, known only up to an arbitrary constant,

which satisfy the law of conservation of energy

$$T + V = \text{const.} \tag{1.2}$$

Generally T,V are differentiable and the positions of equilibrium may be defined as those where $dV = 0$, i.e. where

$$\frac{\partial V}{\partial x_i} = 0 \qquad (i = 1,2,\ldots,n) \tag{1.3}$$

where n is the number of parameters x_i. One may however dissociate the problem altogether from conditions of differentiability by defining a position of equilibrium as an extremum of V. This definition will be ample for our purpose.

Now Lagrange formulated the following theorem first proved (later) by Dirichlet:

(1.4) Whenever in a certain position of the system V is minimum then equilibrium in that position is stable.

The appropriate definition of stability is:

(1.5) Corresponding to any $\varepsilon > 0$ there is an $\eta > 0$ such that if $\|\vec{x}\| + \|\vec{x}'\| < \eta$ at the beginning of the movement, it remains $< \varepsilon$ ever after.

This property will now be proved.

(1.6) For convenience we assume that the minimum of V occurs at $\vec{x} = 0$ and that $V(0) = 0$. Since V is continuous, there is a $\rho > 0$ such that in $\mathcal{S}(0,\rho) = \mathcal{S}(\rho)$:

$\|\vec{x}\| < \rho$, we have $V(\vec{x}) > 0$, save at $\vec{x} = 0$ where $V(0) = 0$. Consider now any $\varepsilon < \rho$. Since V is positive on the compact set $B : \|\vec{x}\| = \varepsilon$, it will have a positive lower bound μ on the set. Similarly T is continuous and positive on the compact product of the sets $\|\vec{x}\| \leq \varepsilon$, $\|\vec{x}'\| = \varepsilon$, and so it has a positive lower bound ν on the product.

Take now Σ to an initial position $(\vec{x}^0, \vec{x}'^0)$ where at all events $\vec{x}^0 \in \mathcal{S}(\varepsilon)$, and denote by T_0, V_0 the corresponding values of T,V. Then

$$(1.7) \qquad T = T_0 + V_0 - V.$$

Now T_0 and $T_0 + V_0$ are ≥ 0 and vanish only when $\vec{x}^0 = \vec{x}'^0 = 0$. Since they are continuous there is an $\eta < \varepsilon$ such that $\|\vec{x}^0\| + \|\vec{x}'^0\| < \eta \longrightarrow T_0 + V_0$ and $T_0 <$ $\inf(\mu, \nu)$. By (1.7) then $V < \mu$, $T < \nu$, hence $\|\vec{x}\| + \|\vec{x}'\| < 2\varepsilon$. This proves Lagrange's theorem.

(1.8) It may be observed that differentiability plays no role whatever in Dirichlet's proof. Indeed the proof does not even make use of the fact that $\vec{x}' = \frac{d\vec{x}}{dt}$; it could be a vector unrelated to $\vec{x}$. Nor did the finiteness of the dimension of the spaces of $\vec{x}, \vec{x}'$ enter into the argument.

Assuming now that we are dealing with a system Σ of a finite degree of freedom, and that T,V possess all the first partials as to their variables, the motion of the system is governed by Lagrange's equations, direct derivatives from Newton's laws:

$$(1.9) \qquad \frac{d}{dt}\left(\frac{\partial T}{\partial x_i'}\right) - \frac{\partial P}{\partial x_i} = - \frac{\partial V}{\partial x_i} .$$

Generally T is a quadratic in the x_i' and so (1.9) is equivalent to a system with constant field. The system will have the solution $\vec{x} = \vec{x}' = 0$ which is a critical point, and stability in the sense of (1.5) asserts an evident property regarding the characteristics passing near the singular point. This property may be formulated for more general systems, and so we may expect a stability notion for differential equations. Our next object will be to give its explicit definition.

2. (2.1) We consider then our usual real system

$$\frac{d\vec{x}}{dt} = \vec{p}(\vec{x},t) \tag{2.2}$$

under the assumptions of the theorem of existence (II, 4.1), save that it is assumed here that the domain of (2.2) in the sense of (II,10) is of the form $L \times R$, where L is the real t line and R a connected open subset of the basic vector space $\mathcal{V}_x$. Let $\mathcal{K}$ be a family of trajectories, $\Gamma_0 : \vec{x}^0(t)$ a trajectory of $\mathcal{K}$.

(2.3) Definition. The trajectory Γ_0 is said to be stable relatively to the family $\mathcal{K}$ of trajectories, whenever given any t^0 and any $\varepsilon > 0$ there is an $\eta > 0$ and a time τ such that if $\vec{x}^1(t) \in \mathcal{K}$ and $\|(\vec{x}^0(t^0) - \vec{x}^1(t^0)\| < \eta$, then for every $t > \tau$ we will have $\|(\vec{x}^0(t) - \vec{x}^1(t)\| < \varepsilon$. The stability is said to be of asymptotic type whenever $\| \vec{x}^0(t) - \vec{x}^1(t) \| \to 0$ when $t \to +\infty$ and this for every $\Gamma \in \mathcal{K}$.

In a somewhat less precise way one may say that if at time t^0 the point of any trajectory Γ of $\mathcal{K}$ passes close enough to the corresponding point of Γ_0, then for every $t > \tau$ it will remain quite close to the corresponding point of Γ_0.

Since the solution $\vec{x}(t)$ is a continuous function of the initial point if we replace t^0 by t^1 and Γ_0 has the property just described relative to t^0 then it will still have it relative to t^1. Thus the property just described is in fact independent of the initial time.

(2.4) Definition. The trajectory Γ_0 is said to be unstable if the only class relative to which it is stable consists of Γ_0 itself.

(2.4.1) A direct characterization is as follows: given any ε and any trajectory $\vec{x}'(t)$ such that

$\| (\vec{x}(t^0)-\vec{x}'(t^0) \| < \varepsilon$, then for some $t^1 > t^0$ we will have $\| (\vec{x}(t')-\vec{x}'(t') \| > \varepsilon$.

(2.5) Definition. The trajectory Γ_0 is said to be absolutely stable, or more simply stable (of asymptotic type or otherwise) whenever it is stable (of asymptotic type or otherwise) relative to all classes $\mathcal{K}$ i.e. whenever it is stable (of asymptotic type or otherwise) relative to the class of all trajectories.

(2.6) Definition. A trajectory Γ_0 which is neither absolutely stable nor unstable is said to be conditionally stable.

Such a trajectory is stable with respect to some partial class $\mathcal{K}$, but not with respect to the class of all trajectories.

Examples of each type will be considered later.

(2.7) As a special case of the preceding situation a critical point $\vec{x}^0$ would be a point of equilibrium of the system. The corresponding definitions are:

(2.8) The critical point M is stable relative to $\mathcal{K}$ whenever given t^0 and any $\varepsilon > 0$ there is an $\eta > 0$ and a τ such that if $x^1(t) \in \mathcal{K}$ and $\vec{x}^1(t^0)$ is in the sphere $\mathcal{S}(M,\eta)$ then for $t > \tau$, $\vec{x}^1(t)$ remains in $\mathcal{S}(M,\varepsilon)$. If this holds for no family $\mathcal{K}$ other than M, the point is unstable; while if it holds for the family of all trajectories M is stable. If M is neither stable nor unstable it is conditionally stable. Stability is defined as being of asymptotic type whenever $\| \vec{x}'(t) \| \to 0$ when $t \to +\infty$.

(2.9) Positive and negative stability. The preceding definitions refer to the behavior of the trajectories for $t > \tau$. To underscore the fact one may prefix the terms "stable, ... " with "positive". Thus one may have critical points which are positively unstable, etc. The same questions may be raised regarding the behavior of the trajectories for $t < \tau$, and they will lead to the concepts of negative stability, ...

• (2.10) An example. Consider a real homogeneous system with constant coefficients

$$\frac{d\vec{x}}{dt} = A\vec{x} \tag{2.10.1}$$

and suppose the characteristic roots of A distinct and none with zero real parts. The transformation of coordinates for the reduction of A to its real canonical form will replace (2.10.1) by a similar system satisfied by the real points

$$\frac{dx_i}{dt} = \lambda_i x_i$$

$$\frac{d\bar{x}_i}{dt} = \bar{\lambda}_i \bar{x}_i$$

$$(i = 1, 2, \ldots, r)$$

$$\frac{dx_{2r+i}}{dt} = \lambda_{2r+i} x_{2r+i}$$

where x_{2r+i} and λ_{2r+i} are real. The real general solution is

$$\vec{x} = (C_1 e^{\lambda_1 t}, \bar{C}_1 e^{\bar{\lambda}_1 t}, \ldots, C_r e^{\lambda_r t},$$

$$\bar{C}_r e^{\bar{\lambda}_r t}, C_{2r+1} e^{\lambda_{2r+1} t}, \ldots, C_n e^{\lambda_n t}).$$

Since $Ce^{\lambda t} \to 0$ for $t \to +\infty$ when the real part of λ is negative, and $Ce^{\lambda t} \to +\infty$ for $t \to +\infty$ when the real part is positive we can assert this: Let k be the number of λ_i with negative real parts. Making all the constants C_j except those corresponding to the λ_i with negative real parts equal to zero, there is obtained a k-fold family of characteristics stable as to the origin. The family is clearly maximal. Hence we have:

conditional stability of asymptotic type for $k < n$;
absolute stability of asymptotic type for $k = n$;
instability for $k = 0$.

As we shall see later, this type of behavior is quite general.

3. (3.1) Uniformly regular transformations. In discussing stability it is sometimes convenient to have transformations which do not affect the stability properties at the origin. Such a type is introduced here.

Consider a mapping $T : \vec{y} = \vec{f}(\vec{x},t)$ of $\mathcal{V}_x$ into $\mathcal{V}_y$ which sends the origin of $\mathcal{V}_x$ into the origin of $\mathcal{V}_y$ and is topological between two fixed neighborhoods, U of the origin in $\mathcal{V}_x$ and V of the origin in $\mathcal{V}_y$ within which we will operate throughout.

We assume the usual properties, notably that the partial derivatives $\frac{\partial f_i}{\partial x_j}$ exist wherever they are needed and that the Jacobian $J = \left|\frac{\partial f_i}{\partial x_j}\right| \neq 0$ for $\vec{x} \in U$ and all $t > \tau$, where τ is fixed once for all.

(3.2) Definition. The mapping T is said to be uniformly regular at the origin whenever given any $\rho > 0$ there is a corresponding $\rho_1 > 0$ such that $\|\vec{x}\| \geq \rho$ and $t > \tau \rightarrow \|T\vec{x}\| \geq \rho_1$ and likewise for T^{-1}.

We prove:

(3.3) A mapping which is uniformly regular at the origin preserves the stability properties in both directions.

It is sufficient to prove that stability asymptotic or otherwise in $\mathcal{V}_y$ implies the same in $\mathcal{V}_x$.

Consider first stability. Denote by $\mathcal{S}(\varepsilon)$, $\mathcal{S}_1(\varepsilon)$ the spheres $\|\vec{x}\| < \varepsilon$, $\|\vec{y}\| < \varepsilon$ in $\mathcal{V}_x$, $\mathcal{V}_y$. We must show that given any $\varepsilon > 0$ there is an $\eta > 0$ such that if $\vec{x} \in \mathcal{S}(\eta)$ at any time $t > \tau$ then $x \in \mathcal{S}(\varepsilon)$ at any ulterior time. Now corresponding to our ε and τ there is an ε_1 such that $0 < \varepsilon_1 < \|T\vec{x}\|$ for $\|\vec{x}\| \geq \varepsilon$ and all $t > \tau$. Hence for all $t > \tau$ the original $T^{-1}\vec{y}$ of any point $\vec{y} \in \mathcal{S}_1(\varepsilon_1)$ lies in $\mathcal{S}(\varepsilon)$. By the assumption of stability in $\mathcal{V}_y$ there is an η_1 such that if $\vec{y} \in \mathcal{S}_1(\eta_1)$ at any time $t > \tau$ then $\vec{y}$

remains in $\mathscr{S}_1(\varepsilon_1)$ ever after. Clearly $T^{-1}\mathscr{S}_1(\eta_1) \subset \mathscr{S}(\varepsilon)$. By our restriction on the mapping there is an $\eta > 0$ such that $\|\vec{y}\| \geq \eta_1 \longrightarrow \|T^{-1}\vec{y}\| > \eta$ for all $t > \tau$. Hence $\bigcap_{t>\tau} T^{-1}\mathscr{S}_1(\eta_1) \supset \mathscr{S}(\eta)$. If $\vec{x} \in \mathscr{S}(\eta)$ then $T\vec{x} \in \mathscr{S}_1(\varepsilon_1)$ for every $t > \tau$ and hence $\vec{x} \in \mathscr{S}(\varepsilon)$ for every $t > \tau$. Thus we have stability in $\mathcal{V}_x$ and so (3.3) holds as regards stability.

Suppose we have now asymptotic stability in $\mathcal{V}_x$. To prove that we have it also in $\mathcal{V}_y$ we must show that if $\varepsilon \longrightarrow 0$ then there is an $\varepsilon_1(\varepsilon) \longrightarrow 0$ with ε such that $T\mathscr{S}(\varepsilon) \subset \mathscr{S}_1(\varepsilon_1)$ for $t > \tau$. If this is false then there is an $\alpha > 0$ such that $T\mathscr{S}(\varepsilon) \cap (\mathcal{V}_y - \mathscr{S}_1(\alpha))$ is non-void for all $\varepsilon > 0$. By hypothesis there is an $\eta > 0$ such that if $t > \tau$ and $\|\vec{y}\| \geq \alpha$ then $\|T^{-1}\vec{y}\| \geq \eta$. Hence if $\varepsilon = \eta/2$ we will have a contradiction and so (3.2) holds likewise regarding asymptotic stability.

(3.4) It is to be noted that we needed the restriction on T <u>both ways</u> in order to prove the stability result going from one space to the other. It is not enough to assume only that $0 < \rho_1 \leq \|T\vec{x}\|$ for $\|\vec{x}\| \geq \rho$ and $t > \tau$.

(3.4.1) As an additional remark, it may be noted that if the mapping T is such that

$$0 < \rho_1 \leq \|T\vec{x}\| \leq \rho_2 \text{ when } \|\vec{x}\| = \rho \text{ for all } t > \tau,$$

and $\rho_2 \longrightarrow 0$ uniformly in t as $\rho \longrightarrow 0$, then this is enough to deduce the stability in $\mathcal{V}_x$ from that in $\mathcal{V}_y$ and conversely.

<u>Proof</u>. Assume stability in $\mathcal{V}_x$. Then, given any $\varepsilon > 0$ there exist ε_1 and ε_2 such that $0 < \varepsilon_1 \leq \|T\vec{x}\| \leq \varepsilon_2$ when $\|\vec{x}\| = \varepsilon$ and $t > \tau$. Stability in $\mathcal{V}_y$ implies that we can find an η_1 corresponding to ε_1, and by the restriction on the mapping there is an η sufficiently small such that for $\|\vec{x}\| = \eta$, $0 < \alpha_1 \leq \|T\vec{x}\| < \alpha_2 < \eta_1$ uniformly in t. It is clear that η, together with ε, establishes stability in $\mathcal{V}_x$. A similar proof establishes stability

in $\mathcal{V}_y$ from that in $\mathcal{V}_x$.

(3.5) The verification of the uniform regularity property in any particular case is not a simple matter. The following type of transformation is however sufficiently general to cover all our requirements. T is given by a relation

$$\vec{y} = A(t)\vec{x} + \vec{a}(\vec{x},t) \tag{3.5.1}$$

where for $t > \tau$: (a) the $a_i(\vec{x},t)$ are power series in the x_j beginning with terms of degree at least two convergent in a fixed toroid $\mathcal{T}$: $|x_i| < A$, and with coefficients continuous in t; (b) A(t) is continuous and has bounded terms and both $|A(t)|$ and $|A^{-1}(t)|$ are bounded. Under the circumstances indeed T^{-1} is represented by a relation of same type

$$\vec{x} = A^{-1}(t)\vec{y} + \vec{b}(\vec{y},t). \tag{3.5.2}$$

(3.6) Examples: A non-singular linear transformation with constant coefficients, or a linear transformation with continuous and bounded periodic coefficients whose determinant $\neq 0$ is of the type here considered.

(3.7) Before proving that (3.5) is uniformly regular at the origin it will be convenient here to replace the norm $\|\vec{x}\|$ by the Euclidean length $\delta\vec{x} = (\sum|x_i|^2)^{1/2}$. We notice that $\zeta = \frac{\|\vec{x}\|}{\delta x}$ is homogeneous and of degree zero in the coordinates. Hence ζ takes all its values on the sphere of radius one: $\delta\vec{x} = 1$. Since this sphere is compact and $\|\vec{x}\| > 0$, ζ is included between two fixed positive numbers. It follows immediately that as regards uniform regularity at the origin we may replace everywhere $\|\vec{x}\|$ by $\delta\vec{x}$.

Suppose now first T, hence also T^{-1} linear, and sc $\vec{a} = \vec{b} = 0$ and set $\delta\vec{x} = r$, $\delta\vec{y} = \rho$, where $\vec{y} = T\vec{x}$. We recall that by Schwartz's inequality, if $\vec{\xi}$, $\vec{\eta}$ are real vectors then

$$(\sum \xi_i \eta_i)^2 \leq \sum \xi_i^2 \cdot \sum \eta_i^2.$$

Hence here

$$|y_i|^2 = |\sum a_{ij} x_j|^2 \leq (\sum |a_{ij}| \cdot |x_j|)^2$$
$$\leq \sum |a_{ij}|^2 \cdot \sum |x_j|^2 = \sum |a_{ij}|^2 \cdot r^2.$$

Let α be an upper bound for the absolute values of the terms of both A and A^{-1}. We have then

$$|y_i| \leq n\alpha^2 r^2$$

and therefore

$$\rho^2 = \sum |y_i|^2 \leq n^2\alpha^2 r^2.$$

Similarly $r^2 \leq n^2\alpha^2\rho^2$. Therefore

$$\frac{1}{n\alpha} \leq \frac{r}{\rho} \leq n\alpha.$$

Hence $\rho \geq \varepsilon \longrightarrow r \geq \frac{\varepsilon}{n\alpha}$, and similarly with r and ρ interchanged. Thus when T is linear it is uniformly regular at the origin.

Suppose now T arbitrary of the type considered. Set $\vec{u} = \frac{1}{r}\vec{x}$. Then

$$\vec{y} = r\{A(t)\vec{u}\ r\vec{a}_1\},$$

where $\| \vec{a}_1 \|$ is bounded. It follows that given any small positive λ, say $< 1/2$, we may choose an $R > 0$ such that when $r < R$ then $\| \vec{y} \| \leq (1+\lambda) \cdot \delta(A\vec{x})$. Moreover we may choose R so that it be similarly related to T^{-1}. Hence assuming $\delta\vec{x}, \delta\vec{y} < R$ we will have this time

$$\frac{1}{n\alpha(1+\lambda)} \leq \frac{r}{\rho} \leq n\alpha(1+\lambda),$$

and the conclusion is again the same. Thus the uniformly regularity of (3.5.1) is proved.

(3.8) Orbital stability. Often Liapounoff stability proves too stringent, or at all events only a weaker type may be established. It will be sufficient to describe it for characteristics. Roughly speaking it asserts that if a characteristic γ passes closely to γ_0 then it remains close to γ_0 ever after. More explicitly γ_0 is orbitally stable relatively to a family κ [absolutely] whenever given any $\varepsilon > 0$ there is a corresponding $\eta > 0$ such that if a characteristic $\gamma \in \kappa$ [if any characteristic γ] passes at t^0 through $\mathcal{S}(\gamma_0, \eta)$ then it remains in $\mathcal{S}(\gamma_0, \varepsilon)$ for all $t > t^0$. If this condition fails then γ_0 is said to be orbitally unstable relative to κ [absolutely].

The following example will show the difference between the two types. Consider a family κ of closed characteristics $\{\gamma_\lambda\}$ depending continuously upon a parameter λ. That is to say the solution $\vec{x}(t-\tau, \lambda)$ representing γ is a continuous function of λ. Then orbital stability is here a consequence of continuity. Let however $T(\lambda)$ be the period of $\vec{x}(t-\tau, \lambda)$ and suppose it non-constant, a case easily realized. Then no matter how small ν it is not possible to maintain $\| \vec{x}(t-\tau, \lambda_0) - \vec{x}(t-\tau, \lambda_0+\nu) \|$ arbitrarily small since in time it will come arbitrarily near to $\| x(t^0-\tau, \lambda_0) - x(t^0+kT(\lambda_0)-\tau, \lambda_0) \|$ where k is any preassigned number between 0 and 1.

§2. A PRELIMINARY LEMMA

4. (4.1) Consider the function

$$\begin{aligned} \varphi(x) &= \left(1 - \frac{x}{A}\right)^{-n} - 1 - n\frac{x}{A} \\ &= \binom{n+1}{2}\left(\frac{x}{A}\right)^2 + \binom{n+2}{3}\left(\frac{x}{A}\right)^3 + \dots, \quad |x| < A, \end{aligned} \tag{4.1.1}$$

and in the power series let x be replaced by the formal power series in ε:

$$y = \varepsilon y_1 + \varepsilon^2 y_2 + \dots .$$

The result ordered as to the powers of ε is

$$(4.1.2)\quad \varphi(y) = \varepsilon^2\varphi_2(y_1) + \varepsilon^3\varphi_3(y_1, y_2) + \cdots .$$

Notice that the substitution $y_m \to k^m y_m$, k any scalar, yields the same result as replacing ε by $k\varepsilon$, and hence

$$(4.1.3)\quad \varphi_m(ky_1, \ldots, k^{m-1}y_{m-1}) = k^m\varphi_m(y_1, \ldots, y_{m-1}).$$

Choose now the sequences of numbers $z_1, z_2, \ldots$ as follows: z_1 is arbitrary, while

$$(4.1.4)\qquad z_m = M\varphi_m(z_1, \ldots, z_{m-1}), \qquad m > 1,$$

where M is a fixed positive constant. Then:

(4.2) <u>Lemma</u>. <u>There is a positive constant</u> $\gamma(M,A)$ <u>such that whenever</u> $|z_1| < \gamma(M,A)$ <u>the series</u>

$$(4.2.1)\qquad z = z_1 + z_2 + \cdots$$

<u>is absolutely convergent</u>. (Liapounoff).

If the series is convergent then we must have

$$z = z_1 + M\{\varphi_2(z_1) + \varphi_3(z_1, z_2) + \cdots\}$$

$$= z_1 + M\varphi(z).$$

Consider now

$$(4.3)\qquad F(z_1, z) = -z + z_1 + M\varphi(z) = 0.$$

Since $F(0,0) = 0$, $\frac{\partial F(0,0)}{\partial z} = -1$, by the implicit function theorem there exists a function $z^*(z_1)$ holomorphic in z_1 at the origin, satisfying (4.3) and such that $z^*(0) = 0$. About $z_1^* = 0$ this function may be represented in the form

$$(4.4)\qquad z^*(z_1) = z_1^* z_1 + z_2^* z_1^2 + \cdots$$

where the series at the right has a radius of convergency

$\gamma(M,A) > 0$. To prove the lemma it is sufficient to show that

$$(4.5)_m \qquad z_m = z_m^* z_1^m.$$

At all events the coefficients z_m^* may be determined by substituting in (4.3) and identifying the powers of z_1. We thus obtain

$$z_1^* = 1; \quad z_m^* = M\varphi_m(z_1^*,\dots,z_{m-1}^*), \qquad m > 1.$$

Hence by the homogeneity property (4.1:3) together with $(4.5)_p$, $p < m$,

$$z_1^* z_1 = z_1; \quad z_m^* z_1^m = M\varphi_m(z_1^* z_1,\dots,z_{m-1}^* z_1^{m-1})$$
$$= M\varphi_m(z_1,\dots,z_{m-1}) = z_m, \quad m > 1.$$

Therefore $z^*(z_1) = z = z_1 + z_2 + \dots$, converges absolutely for $|z_1| < \gamma(M,A)$. This proves the lemma.

§3. SOLUTIONS IN THE NEIGHBORHOOD OF A CRITICAL POINT (FINITE TIME)

5. It is clear that the question of stability can only be decided by investigating the trajectories neighboring a given one. We must discuss above all their behavior in the vicinity of a critical point. The known results along that line are for analytic systems and are in close relation to a very general theorem due to Liapounoff which we shall take up next.

Consider the basic differential system

$$(5.1) \qquad \frac{d\vec{x}}{dt} = P(t) \cdot \vec{x} + \vec{q}(\vec{x},t)$$

under the following assumptions: <u>There is an interval</u> $I : t^1 < t < t^2$ <u>such that</u>:

(5.2) <u>the matrix</u> $P(t)$ <u>is continuous in</u> $\bar{I}$;

(5.3) <u>for every</u> $t \in \bar{I}$ <u>the vector</u> $\vec{q}(\vec{x};t)$ <u>is holomorphic in a certain closed sphere</u> $\mathcal{J}(A)$: $\|\vec{x}\| \leq A$ <u>and</u>

<u>its expansion about the origin begins with terms of degree</u> > 2.

Thus $\vec{x} = 0$ is a critical point of (5.1) in the sense that $\vec{x} = 0$ is a trajectory. We are interested in finding the general behavior of the solutions in the neighborhood of the point. This is the object of the theorem of Liapounoff.

(5.4) Notice that since $\vec{q}(\vec{x},t)$ is continuous on the compact set $\mathcal{S}(A) \times \bar{I}$, $\| \vec{q}(\vec{x},t) \|$ has a finite upper bound M on the set.

6. The method of solution will consist in taking a series

$$\vec{x}(t;\varepsilon) = \varepsilon\vec{x}^1(t) + \varepsilon^2\vec{x}^2(t) + \dots , \tag{6.1}$$

where the $\vec{x}^m$ are continuous and differentiable on I, substituting in (5.1) and choosing the $\vec{x}^m(t)$ so that $\vec{x}(t,\varepsilon)$ formally satisfies (5.1) and that $\vec{x}(t) = \vec{x}(t,1)$, i. e.

$$\vec{x}(t) = x^1(t) + x^2(t) \dots \tag{6.2}$$

is an actual solution of the differential equation. The role of ε will be merely to assign a weight to the $\vec{x}^m$. We might in fact dispense entirely with ε, consider $\vec{x}^m$ and its components as of weight m, substitute in (5.1) and identify the terms of same weight.

Substituting then (6.1) in (5.1) and identifying the powers of ε we find:

$$\frac{dx^1}{dt} = P \cdot \vec{x}^1, \tag{6.3$_1$}$$

$$\frac{d\vec{x}^m}{dt} = P \cdot \vec{x}^m + \vec{r}^m(\vec{x}^1,\dots,\vec{x}^{m-1};t), \qquad m > 1. \tag{6.3$_m$}$$

The solution of $(6.3)_1$, $(6.3)_m$ thus gives rise to a process of successive approximation. As in the fundamental existence theorem (II, 4.1) the question is to prove that the process converges and that the limit (6.2) satisfies (5.1).

(6.4) It is clear that the components r_i^m of $\vec{r}^m$ are polynomials in the components of the vectors $\vec{x}^1, \ldots, \vec{x}^{m-1}$. If we make in $x(t;\varepsilon)$ the substitution $\vec{x}^m(t) \to k^m\vec{x}^m(t)$, the effect is the same as replacing ε by $k\varepsilon$. Hence

$$(6.5) \quad \vec{r}^m(k\vec{x}^1,\ldots,k^{m-1}\vec{x}^{m-1};t) = k^m\vec{r}^m(\vec{x}^1,\ldots,\vec{x}^{m-1};t).$$

This homogeneity property will be required in a moment.

7. Consider first the matrix equation

$$(7.1) \qquad \frac{dY}{dt} = P(t)Y.$$

If Y is a non-singular solution any other is of the form YC, C constant. Since $|Y(t^1)| \neq 0$, $Y^{-1}(t^1)$ exists. Taking $C = Y^{-1}(t^1)$, we will have a non-singular solution which reduces to the identity matrix E for $t = t^1$. For convenience we suppose that Y itself is that solution so that henceforth $Y(t^1) = E$.

(7.2) If $\vec{y}^i(t) = (y_{1i},\ldots,y_{ni})$ then the general solution of $(6.3)_1$ is

$$(7.2.1) \qquad \vec{x}^1 = a_1\vec{y}^1 + \ldots + a_n\vec{y}^n.$$

Leaving for the present the a_i as undefined scalars, in accordance with (III, 5.6, 7.4) we may solve $(6.3)_m$, $m > 1$ recurrently for $t^1 \leq t \leq t^2$ by:

$$(7.2.2) \quad \vec{x}^m = Y\int_{t_1}^{t} Y^{-1} \cdot \vec{r}^m(\vec{x}^1,\ldots,\vec{x}^{m-1};t)\,dt.$$

Since $\vec{x}^m(t^1) = 0$, $m > 1$, if the resulting series (6.2) is shown to be convergent, and to satisfy (5.1), it will represent a solution $\vec{x}(t)$ such that $\vec{x}(t^1) = \vec{x}^1(t^1)$. That is to say it will be a function of t, and the a_i, or again a solution with the same "degree of freedom" (in a certain obvious sense) as the general solution of the linear system (6.3). Referring to (II, 7.12) it will be seen that the solution thus obtained will be itself a general solution.

Our main task is now to show that our process of successive approximation, i.e. the resulting series (6.2) is convergent.

Notice also the following property:

(7.3) x_i^m _is a form_ (_homogeneous polynomial_) _of degree_ m _in_ $a_1, \ldots, a_n$.

Since the r_i^m are polynomials in the components of $\vec{x}^1, \ldots, \vec{x}^{m-1}$, it is clear by recurrence, that they are also polynomials in $a_1, \ldots, a_n$, and hence this holds also for x_i^m. Suppose now (7.3) proved for indices $< m$. If we replace a_i by ka_i, $(i = 1,2,\ldots,n)$, then $\vec{x}^\mu$, $\mu < m$, is replaced by $k^\mu \vec{x}^\mu$, and hence, by (6.5), x_i^m is replaced by $k^m x_i^m$ which proves (7.3).

(7.4) It is clear that as regards the solution $\vec{x}^m$ of $(6.3)_m$, without reference to any ulterior consideration, we could very well replace $\int_{t^1}^{t}$ by $\int_{t'}^{t}$, where t' is any point of $\bar{I}$, not necessarily the same for all values of m. The special choice here made has been for the purpose of obtaining a suitable solution of (6.1).

(7.5) Since $|Y(t)| \neq 0$, $t \in \bar{I}$, the absolute values of the terms of $Y(t)$ and $Y^{-1}(t)$ have an upper bound which we may assume common and equal to α.

(7.6) We will set $a = \sup |a_i|$. As we shall see, it will be necessary to impose an upper bound on a.

(7.7) Since we are primarily interested in the behavior of the trajectories near the origin, we will deliberately confine $\vec{x}(t)$ to the sphere $\mathcal{S}(A)$: $\|\vec{x}\| < A$.

If we set

$$q_i(\vec{x},t) = \sum q_i^{(m)}(t) x_1^{m_1} \ldots x_n^{m_n},$$

$$m_1 + \ldots + m_n = m > 1$$

then referring to (5.3) and (5.4) there follows

$$(7.8)\quad q_i(x_1,\ldots,x_n;t) \ll M\left\{\frac{1}{\prod(1-\frac{x_i}{A})} - 1 - \frac{1}{A}\sum x_n\right\} =$$

$$= Q(x_1,\ldots,x_n) \quad ,$$

where both q_i and Q stand for their expansions in powers of the x_i about the origin. Let the effect, on any number or function, of replacing the q_i by Q be indicated by [ ]. Let also ξ_m denote a common upper found for the $|x_i^m|$ and $\vec{\xi}^m$ the vector whose n components are all equal to ξ_m. Evidently

$$(7.9)\quad |r_i^m(\vec{x}^1,\ldots,\vec{x}^{m-1};t)| < [r_i^m(\vec{\xi}^1,\ldots,\vec{\xi}^{m-1})].$$

On the other hand with φ as in (4.1.1):

$$Q(x,\ldots,x) = M\varphi(x),$$

$$[r_i^m(\vec{\xi}^1,\ldots,\vec{\xi}^{m-1})] = M\varphi_m(\xi_1,\ldots,\xi_{m-1})$$

and therefore

$$(7.10)\quad |r_i^m(\vec{x}^1,\ldots,\vec{x}^{m-1};t)| < M\varphi_m(\xi_1,\ldots,\xi_{m-1}).$$

8. We come now to the convergency proof proper. From (7.2.1) follows

$$(8.1)\qquad x_i^1 = \sum a_h y_{ih},$$

and so by (7.5) and (7.6):

$$(8.2)\qquad \| \vec{x}^1 \| < n^2\alpha a.$$

If $Y^{-1} = \| z_{hk} \| = Z$, we have from (7.2.2):

$$(8.3)\quad x_i^m = \sum y_{ih} \int_{t_1}^{t} z_{hk} r_k^m(\vec{x}^1,\ldots,\vec{x}^{m-1};t)dt.$$

Hence if $|t^2 - t^1| = \tau$, (8.3) yields:

(8.4) $$\| \vec{x}^m \| < n^3\alpha^2 M\tau\varphi_m(\xi_1, \ldots, \xi_{m-1}),$$

and so we may choose in fact

(8.5) $$\begin{aligned} \xi_m &= B\varphi_m(\xi_1, \ldots, \xi_{m-1}) \\ B &= n^3\alpha^2 M\tau. \end{aligned}$$

(8.6) By lemma (4.2) $\xi_1 + \xi_2 + \ldots$ converges for ξ_1 sufficiently small and this uniformly relative to fixed A, B. In view of (8.4) and (8.2) the basic series (6.2) converges likewise whenever a is chosen sufficiently small and the convergence is uniform as to t in $\bar{I}$. Since $\vec{x}(t)$ thus obtained vanishes when every $a_i = 0$, and its components are holomorphic in the a_i at the origin, there is a γ such that a, hence every $|a_i| < \gamma \longrightarrow \vec{x}(t) \in \mathcal{J}(A)$, and also that the $x_i(t)$ are holomorphic in the a_i in $\mathcal{J}(A)$. Since $\vec{x}^m(t)$ satisfies $(6.3)_m$ it is differentiable in $\bar{I}$. Consider now the series of the derivatives. We have

(8.7) $$\frac{d\vec{x}^m}{dt} = P\sum \vec{x}^m + \sum_{m>1} \vec{r}^m(\vec{x}^1, \ldots, \vec{x}^{m-1}; t).$$

Of the two series at the right, since $\sum \vec{x}^m$ is uniformly convergent and P(t) has bounded terms in $\bar{I}$, the first series is uniformly convergent. Regarding the second by (4.9), or else directly

(8.8) $$\| \vec{r}^m(x^1, \ldots, x^{m-1}; t) \| < M\varphi_m(\xi_1, \ldots, \xi_{m-1}).$$

Since $\xi = \xi_1 + \xi_2 + \ldots$ is convergent, by (8.5), $\sum \varphi_m(\xi_1, \ldots, \xi_{m-1})$ is likewise convergent. Hence the second series in (8.7) is likewise uniformly convergent. It follows that the series of the derivatives is uniformly convergent. Hence it represents the derivatives of the series $\sum \vec{x}^m(t)$. Thus (8.7) yields

$$\frac{d\vec{x}}{dt} = P\vec{x} + \sum r^{m}(\vec{x}_1, \ldots, \vec{x}^{m-1}; t)$$
$$= P\vec{x} + \vec{r}(\vec{x}; t), \quad t \in \bar{I}.$$

In other words $\vec{x}(t)$ is a solution of (5.1). Moreover by construction $\vec{x}(t^1)$ coincides with $\vec{x}^1(t^1)$.

(8.9) The linear differential equation

(8.9.1) $$\frac{d\vec{y}}{dt} = P(t) \cdot \vec{y},$$

obtained by neglecting at the right in (5.1) (i.e. in the expressions of the components $\frac{dy_i}{dt}$) the terms of degree > 1, is known as <u>the first approximation</u> for (5.1). What we have shown in substance is that (5.1) possesses a solution whose value for $t = t^1$ is the same as that of any solution of the first approximation which is within a certain neighborhood of the origin in the space $\mathcal{V}_x$.

(8.10) We have already observed (7.3) that x_i^m as determined by (7.2.2) is a homogeneous polynomial of degree m in $a_1, \ldots, a_n$. Thus if $\vec{a} = (a_1, \ldots, a_n)$ we may write

$$\vec{x}(t;\vec{a}) = \vec{x}^1(t;\vec{a}) + \vec{x}^2(t;\vec{a}) + \ldots ,$$

where x_i^m is the collection of the terms of degree m in the expansion of x_i in powers of the a_i about $\vec{a} = 0$. If we consider $\vec{a}$ as a small vector of the "first" order, then $\vec{x}^1(t) + \ldots + \vec{x}^m(t)$ represents $\vec{x}(t)$ to within terms of order $m + 1$ relative to $\vec{a}$, i.e. it is the mth approximation to $\vec{x}$ in the same sense. The first approximation in this sense is still $\vec{x}^1(t)$.

If we combine the results obtained so far we find that we have proved the

(8.11) <u>Theorem of Liapounoff</u>. <u>Consider the differential equation</u>

(8.11.1) $$\frac{d\vec{x}}{dt} = P(t)\vec{x} + \vec{q}(\vec{x};t)$$

<u>where P is continuous for t in a certain closed finite interval $t_1 \leq t \leq t_2$, $\vec{q}(\vec{x};t)$ is holomorphic in $\vec{x}$ in a certain fixed neighborhood N_x of $\vec{x} = 0$ for the same range of t, and begins with second degree terms. The process of successive approximation defined by $(6.3)_m$ yields a solution whose value at $t = t_1$ is the same as that of any preassigned solution $\vec{y}$ of the first approximation equation provided that $\vec{y}(t^1)$ is within a suitable neighborhood of the origin. Moreover the solution $\vec{x}(t)$ remains in N_x for $t \in I$.</u>

§4. SOLUTIONS IN THE NEIGHBORHOOD OF A CRITICAL POINT (INFINITE TIME) FOR SYSTEMS IN WHICH THE TIME DOES NOT FIGURE EXPLICITLY

9. Consider a system:

(9.1) $$\frac{d\vec{x}}{dt} = P\vec{x} + \vec{q}(\vec{x}),$$

where the right hand side is independent of t. More precisely, we suppose that the matrix P is constant while the components $q_i(x) = q_i(x_1, \ldots, x_n)$ are power series convergent in a certain neighborhood of the origin in the vector space $\mathcal{V}_x$ and begin with terms of degree ≥ 2. These assumptions are the same as in (5). It so happens that in certain noteworthy cases considerable additional information may be given regarding the form of the solution in the neighborhood of the origin.

(9.2) We shall have occasion to consider finite sets of numbers $\{\lambda_1, \ldots, \lambda_k\}$, viewed as points in the complex plane possessing one or the other or both of the following two properties:

(9.2.1) <u>Convexity property</u>: the convex closure of the set $\{\lambda_1, \ldots, \lambda_k\}$ does not contain the origin.

(9.2.2) <u>Positive independence property</u>: there exist no relations of the form

$$\lambda_i = \sum m_h \lambda_h,$$

where the m_h are integers ≥ 0 and $\sum m_h > 1$. (This implies in particular that none of the λ_i is zero.)

If the set $\{\lambda_i\}$ has both properties, then it is said to be <u>well behaved</u>.

(9.3) <u>Let the set $\{\lambda_i\}$ be well behaved. If $m_1, \ldots, m_n$ are any integers ≥ 0 such that $m = \sum m_j > 1$, then the numbers</u>

$$(9.3.1) \qquad g = \left| \frac{-\lambda_i + \sum m_j \lambda_j}{-1 + \sum m_j} \right|$$

<u>have a positive lower bound β (independent of i).</u>

It is sufficient to prove this for a given i. Evidently if $\lambda = \inf \{|\lambda_i|\}$:

$$g = \left| \frac{\frac{\sum m_j \lambda_j}{m} - \frac{\lambda_i}{m}}{1 - \frac{1}{m}} \right| > \frac{\lambda - \frac{|\lambda_i|}{m}}{1 - \frac{1}{m}}$$

Let m_0 be the first integer $\geq 2|\lambda_i|/\lambda$ and also > 1. For $m \geq m_0$ we shall then have $g \geq \lambda/2 > 0$. The number of sets $(m_1, \ldots, m_n)$ within $m < m_0$ is finite; in view of (9.2.2) the corresponding g is never zero and so it has likewise a positive lower bound for all values of $m < m_0$. Thus g has a positive lower bound for all values of m.

(9.4) <u>Primitives</u>. In the forthcoming proofs we shall have to consider the primitives of certain exponentials and powers of t or their combinations. We shall agree to choose each time

$$(9.4.1) \qquad \int e^{\lambda t} dt = \frac{1}{\lambda} e^{\lambda t} \qquad (\lambda \neq 0)$$

$$(9.4.2) \qquad \int t^n dt = \int_0^t t^n dt = \frac{t^{n+1}}{n+1}$$

(9.4.3) $$\int \frac{t^n}{n!} e^{\lambda t} dt = \frac{e^{\lambda t}}{n+1} \left(\frac{(\lambda t)^n}{n!} - \frac{(\lambda t)^{n-1}}{(n-1)!} + \dots \right)$$

(n a positive integer, $\lambda \neq 0$)

with their natural consequences for the primitives of polynomials and their products by exponentials $e^{\lambda t}$. Notice that if the real part of λ is $\mathcal{R}(\lambda) < 0$, and n is a positive integer, then

(9.4.4) $$\int t^n e^{\lambda t} dt = \int_{+\infty}^{t} t^n e^{\lambda t} dt,$$

and similarly with t^n replaced by a polynomial.

10. The ground is now prepared for two important expansion theorems, which in turn will yield the stability criteria.

(10.1) Theorem. Consider the differential equation (9.1) in which the time does not figure explicitly and also its first approximation

(10.1.1) $$\frac{d\vec{x}}{dt} = P\vec{x}.$$

Suppose that the set $\{\lambda_i\}$ of the characteristic roots of the matrix P is well behaved and that $P \sim \text{diag.}(\lambda_1, \dots, \lambda_n)$. If $\vec{u}$ is a general solution of (10.1.1) then there exists a general solution of (9.1) of the form

(10.1.2) $$\vec{x} = \vec{u} + \vec{f}(\vec{u}),$$

where $\vec{f}$ is holomorphic in the neighborhood of the origin and begins with terms of degree at least two.

(10.2) By a non-singular linear transformation of coordinates $\vec{x}^* = A\vec{x}$, (9.1) is replaced by a similar equation with $A^{-1}PA \sim P$ in place of A. This transformation will not modify the properties that interest us, and so as regards (10.1) the situation will remain the same. We may thus choose the coordinate system so that in place of (10.1.1) we have a similar system with P in the canonical form diag. $(\lambda_1, \ldots, \lambda_n)$. In order to avoid multiplying notations, we shall assume that (10.1.1) already has that property, i.e., that in (10.1.1) P is canonical.

(10.3) The solution of (9.1) is now constructed as in (6) with the following modification. The first term $\vec{x}^1 = (u_1, \ldots, u_n)$ is the general solution of (10.1.1), hence $\vec{x}^1 = (\alpha_1 e^{\lambda_1 t}, \ldots, \alpha_n e^{\lambda_n t})$. As for $\vec{x}^m$ we take

$$\vec{x}^m = Y \int Y^{-1} \vec{r}^m(\vec{x}^1, \ldots, \vec{x}^{m-1}) dt, \tag{10.4}$$

where the primitives are as specified in (9.4). Evidently the coordinates of $\vec{x}^1$ are homogeneous polynomials of the first degree in the u_j. Suppose that we have shown that the x_i for every $\mu < m$ may be expressed as homogeneous polynomials of degree μ in the u_j. In view of (6.5) $r_i^m(\vec{x}^1, \ldots, \vec{x}^{m-1})$ is a homogeneous polynomial of degree m in the u_j. Thus we shall have

$$\begin{aligned} r_i^m &= \sum_{(m)} \rho_i^{(m)} u_1^{m_1} \cdots u_n^{m_n} \\ &= \sum \rho_i^{(m)} \exp\,(\sum m_j \lambda_j)t \cdot \alpha_1^{m_1} \cdots \alpha_n^{m_n}; \quad m = \sum m_i > 1; \end{aligned} \tag{10.5}$$

and so the ith component of $Y^{-1}\vec{r}^m$ is

$$\sum \rho_i^{(m)} \frac{\alpha_1^{m_1} \cdots \alpha_n^{m_n}}{\alpha_i} \exp\,(-\lambda_i + \sum \lambda_j m_j)t.$$

Its primitive is

$$\sum \rho_i^{(m)} \frac{\alpha_1^{m_1} \cdots \alpha_n^{m_n}}{\alpha_i} \frac{\exp\,(-\lambda_i + \sum \lambda_j m_j)t}{-\lambda_i + \sum m_j \lambda_j}.$$

As a consequence we have

$$(10.6) \qquad x_i^m = \sum \frac{\rho_i^{(m)} u_1^{m_1} \cdots u_n^{m_n}}{-\lambda_i + \sum m_j \lambda_j}.$$

Hence x_i^m is homogeneous of degree m in the u_j. Here again as in (7.8), and in its notations:

$$q_i(x_1,\ldots,x_n) \ll Q(x_1,\ldots,x_n).$$

Then

$$|x_i^m| < \sum \frac{[\rho_i^{(m)}]\, |u_1^{m_1} \cdots u_n^{m_n}|}{|-\lambda_i + \sum m_j \lambda_j|}$$

Hence in view of (9.3)

$$|x_i^m| < \sum \frac{[\rho_i^{(m)}] |u_1^{m_1} \cdots u_n^{m_n}|}{(m-1)\beta} = \frac{[r_i^m(\vec{x}^1,\ldots,\vec{x}^{m-1})]}{(m-1)\beta}, \quad m > 1,$$

and so owing to (7.9) for $[\vec{r}^m]$:

$$|x_i^m| < \frac{M}{(m-1)\beta} \varphi_m(\xi_1,\ldots,\xi_{m-1}), \qquad m > 1.$$

Hence finally

$$(10.7) \qquad \| \vec{x}^m \| < \frac{nM}{(m-1)\beta} \varphi_m(\xi_1,\ldots,\xi_{m-1}), \qquad m > 1.$$

This is essentially the same as (8.4), and so the same argument as in (8) will yield the proof that $\vec{x}(t) = \vec{x}^1(t) + \ldots$ is convergent and that it is a solution. There remains to show that it is a general solution.

(10.8) The solution just obtained assumes the explicit form

$$(10.8.1) \quad x_i = u_i + s_i(u_1,\ldots,u_n) = F_i(u_1,\ldots,u_n)$$

where the s_i are power series beginning with terms of degree > 1 and convergent in a neighborhood of $\vec{u} = 0$.

Consider for the present $\vec{u}$, i.e. the u_i, as independent variables. Since the Jacobian $\left|\frac{\partial x_i}{\partial u_j}\right| = 1$ at $\vec{u} = 0$, the system (10.8.1), or in vector form

$$\vec{x} = \vec{u} + \vec{s}(\vec{u}), \tag{10.8.2}$$

<u>is uniformly regular</u> at the origin. It remains regular for any fixed t, with $\vec{u}$ varying only through $\bar{\alpha}$. Thus the solution written $\vec{x}(t,\vec{\alpha})$ has the property that $\vec{x}(t^0,\bar{\alpha})$ defines a topological mapping between the neighborhoods of the origins in $\mathcal{V}_x$ and $\mathcal{V}_\alpha$. Therefore it is a general solution. This completes the proof of (10.1).

(10.9) <u>Noteworthy Complement</u>. <u>Suppose that a subset</u> $\{\lambda_1,\ldots,\lambda_k\}$, $k < n$, <u>of the set of characteristic roots of</u> P <u>is well behaved and such that</u> $P \sim \operatorname{diag}(\lambda_1,\ldots,\lambda_k,P_1)$, <u>where</u> P_1 <u>is of order</u> $n - k$. <u>Then there is a solution</u> $\vec{x}(t) = \vec{f}(\vec{u})$, <u>where</u> $\vec{u} = (u_1,\ldots,u_k)$ <u>and</u> $\vec{f}$ <u>is holomorphic in</u> $\vec{u}$ <u>in a certain neighborhood of the origin in</u> $\mathcal{V}_u$.

The argument is the same. All that it requires is to replace everywhere the α_{k+i}, hence also the u_{k+i}, by zero.

(10.10) <u>Stability Theorem</u>. <u>Under the conditions of</u> (10.1) <u>the stability properties of</u> (9.1) <u>relative to the origin are the same as those of the equation of the first approximation</u> (10.1.1), <u>i.e. the same as those of</u> (3.10).

This is an immediate consequence of the fact that (10.1.2) is uniformly regular at the origin and reduces (9.1) to (10.1.1).

(10.11) Consider again the situation of (10.9) where the real parts of $\lambda_1, \ldots, \lambda_k$ are negative and those of the λ_{k+i} positive. The solution corresponding to every u_{k+j} zero has its components x_h represented about the origin by a system

$$x_h = f_h(u_1,\ldots,u_k) \tag{10.11.1}$$

where the f_h are holomorphic in a neighborhood of the origin and the Jacobian matrix $\left\| \frac{\partial f_h}{\partial n_j} \right\|$ is of rank k there. The system (10.11.1) represents a neighborhood of a so-called analytical k-manifold about the origin. If M^k denotes the maximal set obtained by analytical extension around the origin then all the stable solutions are in M^k. The set M^k is an analytical manifold in $\mathcal{V}_x$ which may have "self-intersections" (in a certain evident sense). A similar M^{n-k} exists associated with the infinite range to the left $t < \tau$ and "negative stability" and it contains all the unstable solutions. In the neighborhood of the origin the two manifolds have only the origin in common as results at once from (10.11.1) and the analogue for M^{n-k}. To sum up then we have:

(10.12) Under the same conditions as in (10.9), (10.10) the stable solutions are on a certain analytical manifold M^k, the unstable solutions on another analytical manifold M^{n-k}. Both have the origin as a regular point and intersect only at the origin in a certain neighborhood of the latter.

11. (11.1) We shall now take up a proposition in which the canonical form of P is allowed to be absolutely general but by compensation the characteristic roots are restricted in that it is assumed that their real parts are all negative.

(11.2) Theorem. Consider the equation

$$\frac{d\vec{x}}{dt} = P\vec{x} + \vec{q}(\vec{x}) \tag{11.2.1}$$

where P is a constant matrix of degree n, $\vec{q}$ begins with terms of degree at least two and is holomorphic in some closed toroid : $|x_i| \leq A$. Suppose in addition that the characteristic roots λ_i of P all have their real parts $\mathcal{R}\lambda_i < 0$ and satisfy the positive independence property (9.2.2). Let

$$\frac{d\vec{x}}{dt} = P\vec{x}, \tag{11.2.2}$$

the equation of the first approximation, have the base $\{\vec{u}^i(t)\}$. *Then corresponding to the general solution* $\sum a_i\vec{u}^i(t)$ *of* (11.2.2) *there is a general solution of* (11.2.1) *of the form*

$$(11.2.3) \qquad x_i(t) = \sum X_i^{(m)}(t,\vec{a}) \exp t \sum \lambda_i m_i$$

$$m = \sum m_i > 0$$

((m) *sums over partitions of* m) *with the following properties*: (a) $X_i^{(m)}$ *is a polynomial in* t *whose coefficients are forms of degree* m *in* $a_1, \ldots, a_n$, *the terms of degree one making up* $u_1(t)$; (b) *corresponding to any* $\rho > 0$ *there is a time* τ *such that* (11.2.3) *is uniformly convergent for* $\|\vec{a}\| < \rho$ *and* $t > \tau$.

(11.3) Some preliminary observations will first be made. Just as in the case of (10.1) a linear transformation of coordinates does not change the situation and so we assume at the outset the coordinates so chosen that P is canonical: $P = \text{diag}\,(B_1, \ldots, B_r)$. Under the circumstances by (III, 10) there is a non-singular solution of the matrix equation

$$(11.3.1) \qquad \frac{dY}{dt} = PY$$

of the form

$$(11.3.2) \qquad Y = \text{diag}\,(e^{tB_1}, \ldots, e^{tB_r})$$

where $e^{tB_h} = e^{t\lambda_i} Y_h(t)$, Y being the matrix at the right in (III, 10.7). It suffices to know for the present that:

(11.3.3) The terms above the main diagonal in $Y_h(t)$ are zero, those in the main diagonal unity, and the rest of the form $t^k/k!$. As a consequence $|Y_h| = 1$.

(11.3.4) Y^{-1} has the same form as Y with t replaced by -t. Hence

$$Y^{-1} = \text{diag}\,(e^{-t\lambda_1}Y_1(-t),\ldots,e^{-t\lambda_r}Y_r(-t)).$$

(11.4) Let U(t), V(t) be matrices such that

$$U(t) = \text{diag}\,(U_1,\ldots,U_r),\ V(t) = \text{diag}\,(V_1,\ldots,V_r)$$

where U_i, V_i have the same order p_i. Correspondingly any vector $\vec{x}$ may be written

$$\vec{x} = \vec{x}^{(1)} + \ldots + \vec{x}^{(r)}$$

where $\vec{x}^{(i)}$ has the same coordinates of the orders from $p_1 + \ldots + p_{i-1} + 1$ to $p_1 + \ldots + p_i$ as $\vec{x}$ and the others zero. It is then easily shown that a relation

$$\vec{f}(t) = U\int V\vec{g}(t)dt \tag{11.4.1}$$

is equivalent to the system

$$\vec{f}^{i}(t) = U_i\int V_i\vec{g}^{i}(t)dt \quad (i = 1,2,\ldots,r). \tag{11.4.2}$$

12. The ground is now prepared for the proof of (11.2). A general solution $\vec{u}$ of the first approximation (11.2.2) is first chosen and it can be given as

$$u_i = e^{\lambda_i t}\left(a_i + a_{i-1}\frac{t}{1!} + \ldots + a_{i-r}\frac{t^{r_i}}{r_i!}\right). \tag{12.1}$$

Proceeding then as in (10) there is obtained a formal solution

$$\vec{x} = \vec{x}^1 + \vec{x}^2 + \ldots \tag{12.2}$$

where $\vec{x}^1 = \vec{u}$, and $\vec{x}^m$ is defined recurrently, in view of (11.3) and (11.4) by

$$x_i^m = e^{\lambda_i t}\sum y_{ij}(t)\int e^{-\lambda_i t}y_{jk}(-t)r_k^m(\vec{x}^1,\ldots,\vec{x}^{m-1})dt, \tag{12.3}$$

where $y_{ij}(\pm t) = 0$, 1 or $\frac{(\pm t)^s}{s!}$. As for the indefinite integrals, since the integrands are merely sums of products of exponentials and polynomials, the rules of (9.4) apply throughout and provide unique expressions for the x_i^m.

(12.4) By choice x_i^1 is linear in the $a_h e^{\lambda_h t}$ with coefficients polynomials in t. Suppose that it has already been shown that x_i^p, $1 \leq p < m$, is a form of degree p in the $a_h e^{\lambda_h t}$ with coefficients polynomials in t. As a consequence, and in view of (6.5), r_j^m is a similar form of degree m. Hence applying systematically the rules for primitives laid down in (9.4) we find that x_i^m has the same property. Thus this holds for all m, and so $\vec{x}$ as given by (12.2) will have the form (11.2.3). As previously in (8) if we establish the uniform convergence property (11.2b) it will follow that (12.2) is a solution and as in (10.8) it will obviously be a general solution. Thus all that is now required is to show that the uniform convergence property (11.2b) is fulfilled.

13. (13.1) We have just shown in substance that x_i^m has the general form:

$$(13.1.1) \qquad \begin{aligned} x_i^m &= \sum X_i^{(m)} \exp t \sum m_h \lambda_h, \\ m &= \sum m_h > 0, \end{aligned}$$

where the $X_i^{(m)}$ are forms of degree m in $a_1, \ldots, a_n$, with coefficients polynomials in t. Hereafter we shall assume that we have chosen a fixed ρ and specified a corresponding $\tau > 0$ such that

$$(13.1.2) \qquad |a_i| < \rho, \quad i = 1, 2, \ldots, n; \quad t > \tau.$$

The precise choice of τ will be given later.

Under these circumstances we must above all make certain estimates of the $|x_i^m|$ based on (13.1.1) and on the integral representation (12.2). A first difficulty at the outset lies in the fact that while the $X_i^{(m)} \rightarrow \pm\infty$ when $t \rightarrow +\infty$ the exponentials in (13.1.1) $\rightarrow 0$. Similarly as regards the y_{ij}, y_{jk} in (12.3). For this reason it will be necessary to "borrow" as it were from the exponentials suitable factors "cancelling" the growth of the $X_i^{(m)}$ and this will be our first step.

(13.2) Let then $\lambda = \inf |-R\lambda_i| > 0$ and choose $\eta > 0$ such that $4\eta < \lambda$. The quantity 4η is essentially the amount to be borrowed from the λ_i for the purpose already stated.

Let $\mu_i = \lambda_i + 4\eta$, so that still $R\mu_i < 0$. We rewrite the expression of x_i^m given by (13.1.1) in the form:

$$(13.2.1)_m \quad x_i^m = e^{-(m+2)\eta t} \sum \hat{X}_i^{(m)} \exp t \sum m_h \mu_h$$

$$(13.2.2)_m \quad \hat{X}_i^{(m)} = e^{-(3m-2)\eta t} X_i^{(m)}.$$

Since $3m-2 \geq 1$ for $m > 0$ the $\hat{X}_i^{(m)}$ now tend comfortably to zero as $t \to +\infty$ and this is the basic departure from (13.1.1). The other exponential factors are so chosen as to facilitate the various estimates and comparisons later.

(13.3) If we substitute from (13.2.1) in $r_i^m(\vec{x}^1, \ldots, \vec{x}^m)$ it becomes a function of t, written $r_i^m(\vec{x}^1, \ldots, \vec{x}^{m-1})_t$ given by

$$(13.3.1) \quad r_i^m(\vec{x}^1, \ldots, \vec{x}^m)_t = e^{-(m+4)\eta t} \sum e^{-\alpha_s \eta t} R_{is}^{(m)} \exp t \sum m_h \mu_h$$

where $R_{is}^{(m)}$ is a polynomial in the $\hat{X}_h^{(p)}$, $p < m$, without constant term and hence $\to 0$ when $t \to +\infty$.

(13.4) Substituting now from (13.3.1) in the expression (12.3) for x_i^m we obtain

$$(13.4.1) \quad x_i^m = e^{(\mu_i - 3\eta)t} \sum (e^{-\eta t} y_{ij}(t)) \cdot \int (e^{-\eta t} y_{jk}(-t)) R_{ks}^{(m)} \exp t \{(1-m-\alpha_s)\eta - \mu_i + \sum m_h \mu_h\} dt.$$

We have already observed that the $R_{ks}^{(m)} \to 0$ as $t \to +\infty$. Hence the coefficient of $|\ldots|$ in the above integral remains finite as $t \to +\infty$. This makes it possible to replace the indefinite integral by $\int_{+\infty}^{t}$ so that we may now write

$$(13.4.2)_m \qquad x_i^m = e^{(\mu_i - 3\eta)t} \sum (e^{-\eta t} \cdot y_{ij}(t))$$
$$\cdot \int_{+\infty}^{t} (e^{-\eta t} y_{jk}(-t)) R_{ks}^{(m)} \exp t\,|(1-m-\alpha_s)\eta - \mu_i + \sum m_h \mu_h|\,dt.$$

This expression is much more advantageous than (13.3.1) since now in making estimates we shall be able to take certain terms out of the integration sign. However this will come somewhat later.

(13.5) We are now ready to start the estimating process. In so doing we shall compare various linear expressions in the $\exp t \sum m_h \mu_h$ in the sense of $\ll$ as if they were polynomials in the $\exp(\mu_h t)$. Thus if A and B are two such expressions $A \ll B$ will mean that the coefficients of the terms in $\exp t \sum m_h \mu_h$ in B exceed in absolute value those of the corresponding terms in A.

Referring to (12.3) we shall first look for expressions

$$(13.5.1)_m \qquad z^m = e^{-(m+2)\eta t} \sum Z^{(m)}(t) \exp t \sum m_h \mu_h$$

such that $Z^{(m)}(t)$ is real and positive when $t > \tau$ and $\to 0$ as $t \to +\infty$ and such that in addition

$$(13.5.2)_m \qquad x_i^m \ll z^m.$$

Correspondingly we introduce

$$(13.5.3)_m \qquad \zeta^m = e^{-(m+2)\eta t} \sum Z^{(m)}(t)\, |\exp t \sum m_h \mu_h|$$

so that for $t > \tau$:

$$(13.5.4)_m \qquad |z^m| \leq \zeta^m.$$

Conforming also to an earlier notation we shall denote by $\vec{z}^m$, ..., the vector whose components are all equal to z^m, and similarly later in analogous cases.

Consider the following two properties:

Property Π^m. There exists for every $p < m$ a function z^p of the type $(13.5.1)_p$ such that $(13.5.2)_p$ holds for $i = 1, 2, \ldots, n$.

Property Π^∞. There exists for every m a function z^m of the type $(13.5.1)_m$ such that $(13.5.2)_m$ holds for $i = 1, 2, \ldots, n$.

Our major effort will be spent in showing that

(13.6) If Π^m holds for some m then Π^∞ holds also.

Once this result obtained the proof will conclude with little difficulty.

(13.7) Assume that Π^m holds. Let $[r_i^m(\vec{x}^1, \ldots, \vec{x}^{m-1})]$ denote as previously the analogue of the polynomial $r_i^m(\vec{x}^1, \ldots, \vec{x}^{m-1})$ in the indeterminates x_j^p, $p \leq m-1$, when the $q_i(x)$ in the differential equation (11.2.1) are all replaced by their common majorante

$$Q(\vec{x}) = M\left\{\prod\left(1 - \frac{x_i}{A}\right)^{-1} - 1 - \frac{1}{A}\sum x_i\right\}.$$

Since the coefficients in $[r_i^m(\ldots)]$ and likewise the functions $Z^{(m)}(t)$ are all positive we have at once

$$(13.7.1)_m \qquad \begin{aligned} r_i^m(\vec{x}^1, \ldots, \vec{x}^{m-1})_t &\ll [r_i^m(\vec{z}^1, \ldots, \vec{z}^{m-1})] \\ &= e^{-(m+4)\eta t} \sum e^{-\alpha_s t} T_s^{(m)}(t) \exp t \sum m_h \mu_h \end{aligned}$$

where $T_s^{(m)}$ is a polynomial in the $Z^{(p)}$, $p < m$ with real positive coefficients. Hence $T_s^{(m)}$, like $Z^{(m)}$, decreases monotonely to zero as t varies from τ to $+\infty$. By comparing with (13.3.1) we also note that

$$(13.7.2)_m \qquad |R_{is}^m| < T_s^{(m)}.$$

Now if $\mu' = \inf \{-\mathcal{R}\mu_i\}$, $\mu'' = \sup \{-\mathcal{R}\mu_i\}$ so that $\mu', \mu'' > 0$, we have

$$(13.7.3) \qquad \mathcal{R}\{(1-m)\eta-\mu_i+\sum m_h\mu_h\} < (1-m)\eta + \mu'' - m\mu'.$$

Hence there is an m_0 such that whenever $m \geq m_0$ and whatever i the left-hand side < -1. Until further notice we assume $m \geq m_0$.

We shall now impose our first restriction on τ. Since the $y_{ij}(t)$ are polynomials, every $y_{ij}(\pm t)e^{-\eta t} \to 0$ when $t \to +\infty$. Hence we may select τ such that for $t > \tau$:

$$(13.7.4) \qquad |y_{ij}\cdot(\pm t)e^{-\eta t}| < \frac{1}{n}; \quad i,j = 1,2,\ldots,n.$$

Referring now to the integral $(13.4.2)_m$ we find

$$x_i^m \ll e^{(\mu_i-3\eta)t} \sum T_s^{(m)}(t)$$

$$(13.7.5)_m \qquad \times\int_{+\infty}^{t} \exp t\{(1-m-\alpha_s)\eta - \mu_i + \sum m_h\mu_h\}\, dt$$

where we are justified in taking $T_s^{(m)}$ outside the integration sign since it is monotone decreasing and positive in the range under consideration. Since $\{\ldots\}$ has its real part < -1, it is > 1 in absolute value and so removing the integration sign only strengthens $\ll$. Hence remembering $(13.7.1)_m$:

$$x_i^m \ll e^{(\mu_i-3\eta)t} \sum T_s^{(m)} \exp t\{(1-m-\alpha_s)-\mu_i+\sum m_h\mu_h\}$$

$$(13.7.6) \qquad = e^{2\eta t}[r_i^m(\vec{z}^1,\ldots,\vec{z}^{m-1})].$$

Let us set

$$z^m = e^{-(m+2)\eta t} \sum e^{-\alpha_s \eta t} T_s^{(m)} \exp t \sum m_h \mu_h$$
$$= e^{2\eta t}[r_1^m(\vec{z}^1, \ldots, \vec{z}^{m-1})].$$

We have seen that the $T_s^{(m)}$ are positive, monotone decreasing for $t > \tau$ and $\to 0$ when $t \to +\infty$. Hence $Z^{(m)}$ and z^m have the properties specified for them in (13.5). Moreover $(13.7.5)_m$ yields at once $(13.5.2)_m$. Hence if Π^m holds so does Π^∞. This proves (13.6).

(13.8) We shall now select the z^p, $p < m_0$, so as to satisfy Π^{m_0}. Referring to $(13.2.2)_p$, we have whatever p:

$$(13.8.1)_p \quad \hat{X}_1^{(p)} = (b_{10}+b_{11}t+\ldots+b_{1r-1}t^{r-1})e^{-(p+k)\frac{\eta t}{2}}$$

where $k > 0$ and depends solely upon p. If $\beta =$ $\sup \|b_{1h}\|$, then clearly $|\hat{X}_1^{(p)}| < \beta t^r e^{-(p+k)\frac{\eta t}{2}}$ for $t > \tau > 2$. Set

$$(13.8.2)_p \quad Z^{(p)} = e^{-p\frac{\eta t}{2}} Z^{*(p)}, \quad Z^{*(p)} = \beta t^r e^{-k\frac{\eta t}{2}}$$

and accordingly by $(13.5.1)_p$

$$(13.8.3)_p \quad z_p = e^{-(p+2)\eta t} \sum (e^{-(p\frac{\eta t}{2}} Z^{*(p)}) \exp t \sum p_h \mu_h .$$

From

$$\frac{dZ^{*(p)}}{dt} = \beta t^{r-1}(r - \frac{k\eta t}{2})\, e^{-\frac{k\eta t}{2}}$$

follows that if $r > \frac{2r}{k\eta}$ then $Z^{*(p)}$ is monotone decreasing and hence $\to 0$ monotonely. This is then true *a fortiori* for $Z^{(p)}$ also. Let τ be an upper bound > 2 of the finite set of numbers $\frac{2r}{k\eta}$ corresponding to all the $\hat{X}_1^{(p)}$, $p < m_0$, and also large enough to have (13.7.4) hold for $t > \tau$. Then Π^{m_0} holds, and therefore likewise Π^∞

(13.9) Little is left now to bring the proof of our theorem to a close. Let ξ^m, ϕ_m be as in (4) (ξ^m is what was denoted by z_m there) and choose $\xi^1 < \gamma(M,A)$ as prescribed by lemma (4.2) in order that the series

$$(13.9.1) \qquad \xi^1 + \xi^2 + \dots$$

converge. Let also $\bar{\xi}^m$ denote the vector whose n components are all equal to ξ^m.

Define now ζ^p in relation to z^p by $(13.5.3)_p$. That is to say if $-\mu_h'$ is the real part of μ_h, so that $\mu_h' > 0$, set

$$\zeta^p = e^{-(p+2)\eta t} \sum (e^{-\frac{p\eta t}{2}} Z^{*(p)}) \exp(-t \sum p_h \mu_h').$$

Since $Z^{*(p)} \to 0$ monotonely for $t > \tau$ it is clear that likewise $e^{\frac{p\eta t}{2}} \zeta^p \to 0$ monotonely for $t > \tau$. Therefore at the cost of increasing eventually τ, we may choose it such that

$$(13.9.2)_p \qquad \zeta^p < e^{-\frac{p\eta t}{2}} \xi^p \text{ for } p < m_0.$$

Suppose now as before $m \geq m_0$ and as we may well assume $m_0 \geq 2$; assume also that $(13.9.2)_p$ holds for every $p < m$. Recalling that

$$(13.9.3)_m \quad [r_1^m(\bar{\zeta}^1, \dots, \bar{\zeta}^{m-1})] = M\phi_m(\zeta^1, \dots, \zeta^{m-1})$$

and by virtue of the homogeneity property of the ϕ_m (4.1.3), we find from $(13.7.4)_m$ and $(13.7.5)_m$

$$|x_1^m| \leq [r_1^m(\bar{\zeta}^1, \dots, \bar{\zeta}^{m-1})] = e^{2\eta t} M\phi_m(\zeta_1, \dots, \zeta^{m-1})$$

$$\leq e^{(2-\frac{m}{2})\eta t} M\phi_m(\xi_1, \dots, \xi_{m-1}) \leq M\phi_m(\xi_1, \dots, \xi_{m-1}) = \xi_m.$$

Thus for $m \geq m_0$ and with $t > \tau$ and every $|a_i| < \rho$, the $|x_i^m|$ are at most equal to the terms of the convergent series (13.9.1). This proves the uniform convergence of the solution (12.2) and hence also theorem (11.2).

14. Generalization. Stability properties. Let now merely k ($k \leq n$) roots λ_i, say the first k, have real parts $\mathcal{R}\lambda_i < 0$ and satisfy the property (9.2.2). Then in the base for (11.2.2) with P canonical the element $\vec{u}_i$, $i \leq k$, has its components of the form (polynomial in t) $Xe^{\lambda_i t}$. If we take $\vec{x}^1 = \sum a_i \vec{u}^i$, we may reason as before and prove:

(14.1) If the k characteristic roots $\lambda_1, \ldots, \lambda_k$ ($k \leq n$) have negative real parts and satisfy (9.2.2) there still exists a solution of the form (11.2.3) with the same properties as before save that now $i \leq k$, and the $X_i^{(m)}$ are forms in $a_1, \ldots, a_k$.

(14.2) Application to stability. The general solution $\vec{x}(t,\vec{a})$ given by (11.2.3) is a series uniformly convergent for $t > \tau$ whose terms are continuous functions of $\vec{a}$ for $|a_i| < \rho$, $i = 1, 2, \ldots, n$, and vanish for $\vec{a} = 0$. From this uniform convergence we deduce that given any $\varepsilon > 0$ there is an $\eta > 0$ such that $\|\vec{a}\| < \eta \Longrightarrow \|\vec{x}(t,\vec{a})\| < \varepsilon$ for every $t > \tau$.

Consider now any $t^0 > \tau$ and any $\vec{x}^0$ such that every $|x_i^0| < A$. The solution of the family $\{\vec{x}(t,\vec{a}) \mid |a_i| < \rho\}$, if any exists, passing through $\vec{x}^0$ at time t^0 is defined by the relations

$$(14.2.1) \qquad x_i^0 = \sum a_h u_{ih}(t^0) + \ldots$$

where the terms written represent $\vec{x}^1(t^0)$ (the value of the first term in the solution (12.2) for $t = t^0$).

The matrix $\| u_{ih}(t) \|$ is merely a non-singular matrix solution of (11.3.1) and so (III, (3.4), (4.6)):

$$\Delta = |u_{ih}(t^0)| \neq 0.$$

Since the Jacobian

$$\left| \frac{\partial x_i^0}{\partial a_h} \right| = \Delta ,$$

the system (14.2.1) has a unique solution $\vec{a}(\vec{x}^0)$ such that $\vec{a}(0) = 0$, and it is holomorphic in a certain neighborhood of the origin in $\mathcal{U}_x$. Hence given $\eta_1 > 0$ there is a $\eta > 0$ such that $\| \vec{x}^0 \| < \eta \longrightarrow \| \vec{a} \| > \eta_1 \longrightarrow \| \vec{x}(t,\vec{a}) \| < \varepsilon$ for $t > \tau$. Therefore the motion is stable. Moreover since every $\vec{x}^m(t,\vec{a}) \longrightarrow 0$ with increasing t, and the series (11.2.3) is uniformly convergent for $t > \tau$, $\vec{x}(t,\vec{a})$ itself $\longrightarrow 0$ under the same circumstances and so the stability is asymptotic.

(14.2.2) If there are only $0 < k < n$ characteristic roots say $\lambda_1, \ldots, \lambda_k$ with negative real parts, the reasoning proceeds substantially the same way save that we now choose $a_{k+1} = \ldots = a_n = 0$ and in (14.2.1) consider only the first k equations, but otherwise there are only unimportant modifications. Indeed the special set just considered is asymptotically stable. However, if say $\mathcal{R}\lambda_n > 0$ then there is a solution of the type of (10.9) and as in (10.10) it is proved unstable. Thus there is an unstable solution and so in this case we only have conditional stability.

If $k = 0$, i.e. if every $\mathcal{R}\lambda_i > 0$ there are certainly unstable solutions, but whether or not all solutions are now unstable remains, we believe, an open question.

To sum up we have then:

(14.3) <u>Theorem</u>. <u>If all the λ_i have negative real parts and satisfy (9.2.2) the system is asymptotically stable at the origin. If only $k < n$ have that property there is a k-dimensional system of asymptotically stable solutions and certainly some unstable solutions so that we have at best conditional stability.</u>

(14.4) Assuming that $\mathcal{R}\lambda_i < 0$, $i \leq k$, and $\mathcal{R}\lambda_{k+i} > 0$, we can prove as in (10) the analogue of (10.12), with the sole difference that M^{n-k} will be the carrier of the negatively stable solutions. We cannot assert, however, that these negatively stable solutions are positively unstable.

§5. CRITICAL POINTS WHEN THE COEFFICIENTS ARE PERIODIC

15. This particular situation is chiefly of interest as a preparation to the treatment of periodic solutions. Consider then a system of the form (5.1) save that the coefficients of the products of the x_i are now allowed to be periodic functions of t with the same period, assumed for convenience to be 2π. We suppose the equation in the form

$$\frac{d\vec{x}}{dt} = P(t)\vec{x} + \vec{q}(\vec{x};t) \tag{15.1}$$

$$q_j(\vec{x};t) = \sum q_j^{(m)}(t)\, x_1^{m_1} \cdots x_n^{m} \tag{15.2}$$

$$m = \sum m_j > 1$$

where $P = \| p_{jk}(t) \|$ and the $q_j^{(m)}$ have the period 2π. We assume explicitly:

(15.3) The $p_{jk}(t)$ and $q_j^{(m)}(t)$ admit absolutely convergent expansions in Fourier series on $0 \leq t \leq 2\pi$.

As a consequence:

(15.4) The products of the p_{jk}, $q_j^{(m)}$ admit likewise absolutely convergent expansions in Fourier series on the same range.

We will now distinguish two cases according as P is or is not constant.

(15.5) Case I. The matrix P is constant. This will be subdivided into two cases respectively analogous to those of Theorems (10.1) and (11.2), depending on the nature of the characteristic roots $\lambda_1, \ldots, \lambda_n$.

IA. (15.5.1) The two sets $\{\lambda_1, \ldots, \lambda_n, +2\pi i\}$ and $\{\lambda_1, \ldots, \lambda_n, -2\pi i\}$ are both well behaved.

As in (9.3) we prove:

(15.5.2) *If* $m_1, \ldots, m_n, m'$ *is any set of integers such that the* $m_h \geq 0$, $\sum m_h = m > 1$, *then the numbers*

$$\varepsilon = \left| \frac{-\lambda_j + \sum m_h \lambda_h + 2m'\pi i}{-1 + \sum m_h + m'} \right|$$

have a positive lower bound β *independent of* j.

Treating each of the sets $(\lambda_1, \ldots, \lambda_n, 2\pi i)$, $(\lambda_1, \ldots, \lambda_n, -2\pi i)$ as in (9.3) there are obtained two lower bounds β', β'' and the least of the two is a suitable bound β.

As in (10) we may assume that $P = \operatorname{diag}(\lambda_1, \ldots, \lambda_n)$ with (10.1.1) the same first approximation as before. With $u_1, \ldots, u_n$ the same as in (10) we prove:

(15.5.3) *Under our assumptions the general solution* $\vec{x}(t)$ *may be represented in the vicinity of* $\vec{x} = 0$ *in the form* $\vec{x}_1(t)$ = *power series in the* u_i *with coefficients periodic in* t *and of period* 2π, *the range of validity being a certain neighborhood of the origin in the space* $\mathcal{V}_u$.

The argument proceeds as in (10) with certain minor modifications. The x_j^m are first shown to be homogeneous polynomials of degree m in the u_h with coefficients of period 2π, which owing to (15.4), may be expanded in Fourier series on $0 \leq t \leq 2\pi$.

IB. (15.5.4) *The real parts of the* λ_j *are all negative and* (9.2.2) *holds*. The proof of (11.2) is valid without modification. The $X_i^{(m)}$ will now be polynomials in t with bounded periodic coefficients and (11.2) holds with the minor change just indicated in the $X_i^{(m)}$. The stability properties are the same as in (14.2).

(15.6) *Case II. The matrix* P *is periodic*. This case will be reduced to the preceding one by an appropriate transformation of variables. Consider the auxiliary matrix equation

(15.6.1) $$\frac{dX}{dt} = P(t)X,$$

and let X be a non-singular solution. Referring to (III, 11.3) there is a constant non-singular matrix C such that $X(t+2\pi) = X(t)C$, and we may choose X (see III, 11.4) such that C is in canonical form. Let $\mu_1, \ldots, \mu_n$ be the characteristic roots of C. We call them the <u>characteristic exponents</u> of the differential equation (15.6.1).

Referring to (III, 11.5) there exists a matrix B such that $C = e^{2\pi B}$ (the B loc. cit. may clearly be replaced by $2\pi B$) and if λ_j are the characteristic roots of B then the μ_j are the numbers $e^{2\pi\lambda_j}$. In particular $\mathcal{R}\lambda_j = \frac{1}{2\pi}\log|\mu_j|$. Thus $\mathcal{R}\lambda_j < 0$ is equivalent to: $|\mu_j| < 1$, and (9.2.2) to

(15.6.2) There are no relations $\mu_i = \prod \mu_j^{m_j}$, $m_j > 0$ $\sum m_j > 1$.

We now introduce as in (III, 11.8) the periodic matrix
$$Z(t) = e^{tB} \cdot X^{-1}$$

whose period is 2π and the related transformation of variables (III, 11.11) which we write here in vector form:

(15.6.3) $$\vec{y} = Z(t)\vec{x}.$$

We recall that X^{-1} satisfies the adjoint equation to (15.6.1) and so
$$\frac{dX^{-1}}{dt} = -X^{-1} P.$$

Hence differentiating we find:
$$\begin{aligned}\frac{d\vec{y}}{dt} &= e^{tB} \cdot X^{-1} \frac{d\vec{x}}{dt} + \frac{d}{dt}(e^{tB} \cdot X^{-1}) \cdot \vec{x} \\ &= e^{tB} \cdot X^{-1} \cdot (P\vec{x} + \vec{q}(\vec{x};t)) \\ &\quad + B \cdot e^{tB} \cdot X^{-1} \cdot \vec{x} - e^{tB} \cdot X^{-1} \cdot P\vec{x} \\ &= B\vec{y} + e^{tB} X^{-1} \vec{q}(Z^{-1}\vec{y};t)\end{aligned}$$

or finally $\vec{y}$ satisfies the equation:

$$(15.6.4) \qquad \begin{aligned} \frac{d\vec{y}}{dt} &= B\vec{y} + \vec{r}(\vec{y};t) \\ \vec{r}(\vec{y};t) &= Z(t) \,.\, \vec{q}(Z^{-1}\vec{y};t). \end{aligned}$$

It is clear that (15.6.4) falls under Case I. Thus:

(15.6.5) <u>By the transformation</u> (15.6.3) <u>Case</u> II <u>is reducible to Case</u> I.

Since $|Z(t)| \neq 0$, and has bounded terms for all t, the transformation (15.6.3) is uniformly regular at the origin and so the stability properties of the initial equation (15.6.1) and of (15.6.4) are the same. Those of (15.6.4) depend upon the number of $\mathcal{R}\lambda_j$ which are negative, hence upon the number of $|\mu_j|$ which are < 1. Combining with (14.2) we obtain:

(15.6.6) <u>If the characteristic exponents</u> μ_j <u>are such that every</u> $|\mu_j| < 1$ <u>and</u> (15.6.2) <u>holds, the system</u> (15.6.1) <u>is asymptotically stable at the origin. If this is only true for</u> $k < n$ <u>of the</u> μ_j, <u>there is a</u> k-<u>dimensional system of solutions asymptotically stable at the origin and some unstable solutions so that the stability is at best conditional</u>.

§6. STABILITY OF PERIODIC SOLUTIONS

16. Consider the system

$$(16.1) \qquad \frac{d\vec{x}}{dt} = \vec{p}(\vec{x},t)$$

and suppose that there is a known periodic solution $\vec{y}(t)$ whose behavior as to stability is to be ascertained. We shall assume again for simplicity that the period is 2π. Setting $\vec{x} = \vec{y}(t) + \vec{\xi}(t)$, we shall further suppose that for $\|\vec{\xi}\|$ sufficiently small and t arbitrary the

$p_j(\vec{y}+\vec{\xi},t)$ may be expanded in a power series in the ξ_j. Substituting in (16.1) there is obtained a differential equation in $\vec{\xi}$ of the form

$$\frac{d\vec{\xi}}{dt} = P(t)\vec{\xi} + \vec{q}(\vec{\xi};t) \tag{16.2}$$

of the type (15.1) with the origin as critical point. The matrix

$$P = \left| \frac{\partial p_j(\vec{x})}{\partial x_k} \right|_{\vec{x} = \vec{y}(t)}$$

is merely the Jacobian of the p_j as to the x_k taken on the known periodic trajectory. A natural assumption is that $P \neq 0$ and is non-constant. We are thus under Case II. There will be a set of characteristic exponents $\mu_1, \ldots, \mu_n$, and we may now state as a consequence of (15.6.6):

(16.3) <u>Theorem</u>. <u>The stability properties are the same as those of</u> (15.6.6).

17. <u>Remark regarding periodic motions in a system in which the time does not figure explicitly</u>. Consider the system

$$\frac{d\vec{x}}{dt} = \vec{p}(\vec{x}) \tag{17.1}$$

and let $\vec{y}(t)$ be a known periodic solution which we assume to be of period 2π. We form again (16.2) with the associate characteristic exponents $\lambda_1, \ldots, \lambda_n$. Unfortunately this time (16.3) cannot be applied, as the λ_j do not fulfill (15.2.1). In fact:

(17.2) <u>One of the characteristic exponents</u> λ_j <u>is zero</u> (Poincaré).

We notice in the first place that

$$\frac{d\vec{\xi}}{dt} = P(t)\vec{\xi} \qquad P(t) = \left\| \frac{\partial p_i}{\partial x_j} \right\| \tag{17.3}$$

is the variation equation for the solution $\vec{y}(t)$ of (17.1). Since we are dealing with a constant field $\vec{y}(t+\tau)$ is a solution for all τ. Since we have an analytical situation by (III,3.8): $\vec{\xi}(t) = \frac{\partial \vec{y}(t+\tau)}{\partial \tau} = \frac{d\vec{y}}{dt} = \vec{p}(\vec{y}(t+\tau))$ is a solution and it is periodic of period 2π. Thus the linear transformation θ of the space of the solutions of (17.3) induced by $t \to t + 2\pi$, is such that a certain solution is multiplied by unity. Since the characteristic exponents λ_j include all numbers by which a solution may be multiplied as a consequence of θ, one of the λ_j is unity.

(17.3) Thus in an obviously important case we are faced with an exception to our stability conditions. To deal with the problem an entirely different argument will be required (see (19)).

18. <u>Periodic motions in the vicinity of a given periodic motion</u>. Consider a general system such as (II, 9.1.1)

$$\frac{d\vec{x}}{dt} = \vec{p}(\vec{x},t,\vec{y}) \tag{18.1}$$

and with the same analytical properties as described in (II, 9.1) save that in addition the solution $\vec{\xi}(t)$ is periodic with a period T and exists for all values $t^0 \leq t \leq t + T$, hence for all $t \geq t^0$. If we set $\vec{x}^0 - \vec{\xi}(t^0) = \vec{\eta}$, then by (II,9.1) there is a general solution $\vec{x}(t,\vec{\eta},\vec{y})$ valid in a certain toroid $\mathcal{T}$: $|\eta_i| < a$, $|y_i| < a$, for all t in some interval containing the set $t^0 \leq t \leq t + T$. The condition that the solution possess the period T is

$$-\vec{x}(t^0,\vec{\eta},\vec{y}) + \vec{x}(t^0+T,\vec{\eta},\vec{y}) = 0, \tag{18.2}$$

or more explicitly

$$x_i(t^0+T,\vec{\eta},\vec{y}) - x_i(t^0,\vec{\eta},\vec{y}) = 0. \tag{18.3}$$

This is a system of n analytical equations in the unknowns $\eta_1, \ldots, \eta_n$. If the Jacobian J of the left-hand sides as to the η_h does not vanish for $\vec{\eta} = \vec{y} = 0$, then (18.3) may be solved for $\vec{\eta}$ as a function $\vec{\eta}(\vec{y})$ holomorphic in a neighborhood of the origin with $\vec{\eta}(0) = 0$. The solutions thus defined will form an analytical family containing the given periodic solution $\vec{\xi}(t)$, which is in fact the solution of the family corresponding to $\vec{y} = 0$.

(18.4) When $\vec{p}(\vec{x},t,\vec{y}) = p(x,\vec{y})$, i.e. when the field is constant, an important simplification is possible. We assume that the characteristic Γ_0 corresponding to $\vec{\xi}$ is not a critical point. Hence if $\vec{x}^0 \in \Gamma_0$ one of the components of $\vec{p}(\vec{x}^0,0)$ is $\neq 0$. Let the coordinates be so chosen that $p_n(\vec{x}^0,0) \neq 0$. We may view $\vec{x}^0$ as the initial point of Γ_0 for the time $t = 0$. Let $\vec{x}(t,\vec{z},\vec{y})$ be the solution of (18.1) with the initial value $\vec{z}$ for $t = 0$, as given by (II, 9.1), so that it is here analytical in $t,\vec{z},\vec{y}$ about $(\vec{x}^0,0)$ for all t. Consider the equation in t:

(18.4.1) $$x_n(t,\vec{z},\vec{y}) = x_n^0.$$

Since

$$\left(\frac{\partial x_n}{\partial t}\right)(0,\vec{x}^0,0) = p_n(\vec{x}^0,0) \neq 0,$$

there is a unique analytical solution $t^*(\vec{z},\vec{y})$ of (18.4.1) such that $t^*(\vec{x}^0,0) = 0$. Correspondingly the characteristic Γ_y for $\vec{x}(t,\vec{z},\vec{y})$ intersects $x_n = x_n^0$ in a single point $\to \vec{x}^0$ when $\vec{z} \to \vec{x}^0$ and $\vec{y} \to 0$. Hence we may take as initial point of Γ_y at time $t = 0$ a point whose last coordinate is x_n^0, i.e. assume that in $\vec{z}$ the last coordinate z_n is fixed and equal to x_n^0. Thus $\vec{x}(t,\vec{z},\vec{y})$ will be analytical in $z_1, \ldots, z_{n-1}$ and $\vec{y}$. The corresponding general solution will be $\vec{x}(t-\tau, z_1,\ldots,z_{n-1},\vec{y})$.

(18.4.2) In the present instance there is of course no reason to assume the period to be fixed and equal to

T. Since we are looking for a closed characteristic $\to \Gamma_0$ when $\vec{y} \to 0$, its period will be $T + \tau(\vec{y})$, where $\tau(\vec{y}) \to 0$ with $\vec{y}$. The analytical system will be

$$(18.4.3) \quad x_i(T+\tau, z_1, \ldots, z_{n-1}, \vec{y}) = x_i(0, z_1, \ldots, z_{n-1}, \vec{y})$$

$$(i = 1, 2, \ldots, n).$$

If the Jacobian

$$\left| \frac{\partial[x_i(t, z_1, \ldots, z_{n-1}, 0) - x_i(0, z_1, \ldots, 0)]}{\partial(t, z_1, \ldots, z_{n-1})} \right| \neq 0$$

for $t = T$ and $z_i = x_i^0$ $(i = 1, 2, \ldots, n-1)$, then there is a family of closed characteristics $\{\Gamma_y\}$ such that $\Gamma_y = \Gamma_0$ for $\vec{y} = 0$. However this is a sufficient but not a necessary condition.

19. Let us return to the question raised in (17.3) regarding stability in a system in which the time does not figure explicitly. Take a closed characteristic Γ : $y(t)$ of (17.1) and let $\{\mu_i\}$ be its characteristic exponents with $\mu_n = 1$. Thus the characteristic numbers are the numbers $\lambda_i = \log \mu_i$; and $\lambda_n = 0$. We shall prove:

(19.1) <u>Property (15.6.6) still holds with n replaced by n - 1, i. e. the set $\{\mu_i\}$, $i < n$, now plays the same role as the full set of characteristic exponents did before.</u>

(19.2) Assume first that : (a) $|\mu_i| < 1$ for $i < n$; (b) $\mu_1, \ldots, \mu_{n-1}$ satisfy (15.6.2). In other words $\{\lambda_1, \ldots, \lambda_{n-1}\}$ is a well behaved set. Applying now (11.2) and (15.6.6) to the system (16.2) associated with (17.1), we find that (16.2) has a solution $\vec{\xi}(t, a_1, \ldots, a_{n-1}) = \vec{\xi}(t, \vec{a})$ holomorphic in $\vec{a}$ in a certain neighborhood of the origin in $\mathcal{V}_a$, where $\vec{\xi} \to 0$ with $\vec{a}$. Moreover $\vec{\xi}$ is such that the Jacobian matrix

$\left\| \frac{\partial \xi_i}{\partial a_j} \right\|$ is of rank n - 1 for $\vec{a} = 0$. Finally the

$$\vec{x}(t,\vec{a}) = \vec{y}(t) + \vec{\xi}(t,\vec{a})$$ represents a family $\mathcal{F}$ of solutions of (17.1) containing Γ and in which Γ is asymptotically stable. All this is implicit in the form of the solution $\vec{\xi}(t,\vec{a})$ of (16.2) resulting from (11.2).

(19.3) It is clear that $\left\| \frac{\partial x_i}{\partial a_j} \right\|$ is likewise of rank n - 1 for $\vec{a} = 0$. We may assume therefore that for $\vec{a} = 0$ the determinant $\left| \frac{\partial x_i}{\partial a_j} \right| \neq 0$, $i = 1, 2, \ldots, n - 1$. Let us take some point M of Γ as the origin and suppose that the space $\Pi : x_n = 0$ is tangent to Γ at M. If so a suitable transformation of coordinates

$$x_i' = x_i, \ i < n; \qquad x_n' = \sum k_i x_i, \ k_n = 1$$

will eliminate this difficulty. We may suppose therefore Π non-tangent to Γ at M. Under the circumstances M is an interior point for the set a of all the intersections of the characteristics of the family $\mathcal{F}$ with Π. Moreover, every characteristic Δ of (17.1) passing near enough to M will intersect Π in a point of A. It follows that $\Delta \in \mathcal{F}$, and so Γ is stable in the family of <u>all</u> the characteristics, i. e. it is absolutely stable. Thus (19.1) holds in the present instance. When less than n - 1 of the μ_i satisfy the properties (a), (b) of (19.2), (19.1) follows from (15.6.6). Thus (19.1) is proved.

CHAPTER V

TWO DIMENSIONAL SYSTEMS

§1. GENERALITIES

1. The systems to be investigated in the present Chapter are real systems of the form

$$(1.1) \qquad \frac{dx}{dt} = P(x,y), \quad \frac{dy}{dt} = Q(x,y)$$

where P, Q are real entire functions. It will be assumed that the critical points, the solutions of $P = Q = 0$, are isolated. Thus their number in any bounded region is finite. The central topics discussed are the simpler critical points, the index and its applications and the general aspect of the characteristics in the large. The basic mémoires are those of Poincaré already quoted and of Bendixson (Acta Mat. 24 (1901)).

(1.2) Let O be a critical point. Without changing the form of (1.1) one may choose O as the origin. Then in the neighborhood of O we have

$$(1.3) \qquad P = ax + by + P_2(x,y), \qquad Q = cx + dy + Q_2(x,y)$$

where in this chapter designations such as $P_2(x)$, $Q_3(x,y)$, ..., will stand for power series convergent in some neighborhood of the origin and beginning with terms of degree at least two, three, ...

The characteristic equation of the system (1.3) is

$$(1.4) \qquad (a-\lambda)(d-\lambda) - bc = 0.$$

We shall only consider the case when the characteristic roots λ_1, λ_2 are $\neq 0$, i.e. where $ad - bc \neq 0$. The corresponding critical points are said to be <u>elementary</u>.

We shall show that the topological character of the critical point is essentially determined by its character for the first approximation, i.e. by the system obtained upon replacing P, Q by their first degree terms. This system is merely a linear homogeneous system with constant coefficients and so we shall first discuss such systems.

(1.5) We shall have repeated occasion to utilize analytic transformations of the form

$$(1.5.1) \qquad x = f_1(u,v), \quad y = g_1(u,v)$$

which are regular at the origin, that is to say such that the Jacobian

$$J = \frac{D(f_1,g_1)}{D(u,v)} \neq 0 \text{ for } u = v = 0.$$

We recall the following properties. Let S^{-1} denote the transformation from the (x,y) plane Π to the (u,v) plane Ω. Then:

(1.5.2) There exist two neighborhoods M, N, of the origins in Π and Ω which are 2-cells, and such that S maps N topologically onto M.

(1.5.3) S is regular in N and S^{-1} is regular in M.

It is also an elementary matter to prove:

(1.5.4) S^{-1} transforms the system (1.1) into a similar system and so that characteristics, critical points, isolated critical points of (1.1) go into the same for the new system. Moreover S^{-1} does not change the canonical matrix to which $\begin{Vmatrix} a & b \\ c & d \end{Vmatrix}$ may be reduced.

These properties mean in substance that we shall be free to use transformations such as S without affecting the behavior of the characteristics about the origin.

(1.6) Characteristic Rectangle. We recall that according to (II, 14.3, 14.6) if γ is a characteristic (not a critical point) and A is a point of γ, then a certain neighborhood of A may be mapped topologically on a rectangle so as to produce the configuration of fig. 1

(for convenience the characteristics and their arcs are identified with their topological images) where BC, B'C' are arcs of characteristics and EH, FG are subarcs of preassigned arcs transverse to γ at B, C. Moreover every characteristic passing sufficiently near A contains an arc such as B'C'. The dotted rectangle EFGH is a <u>characteristic rectangle</u> and the arc BC of γ is its axis.

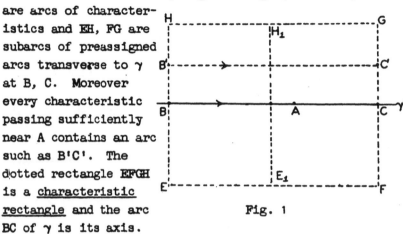

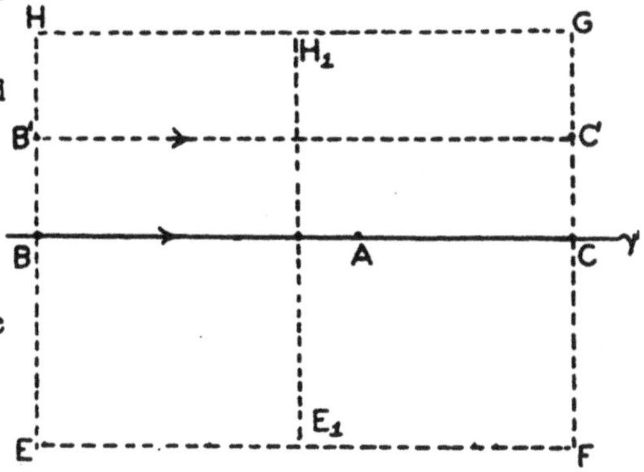

Fig. 1

(1.7) Closed Characteristics. According to (II, 15) the characteristic γ is closed whenever it is a Jordan curve containing no critical point or equivalently whenever its solution $(x(t), y(t))$ is periodic.

§2. LINEAR HOMOGENEOUS SYSTEMS

2. Consider the real system

$$(2.1)\quad \frac{dx}{dt} = ax + by,\ \frac{dy}{dt} = cx + dy,\ ad - bc \neq 0.$$

The characteristic roots λ_1, λ_2 are still the solutions of (1.4) and the reduced canonical form of the coefficient matrix $\left\| \begin{matrix} ab \\ cd \end{matrix} \right\|$ will completely determine the behavior of the characteristics, and more particularly their behavior in the neighborhood of the critical point at the origin. If the characteristic roots are real then a real transformation of coordinates will reduce the system to one of the same form but with coefficient matrix of one of the two types

$$A : \left\| \begin{matrix} \lambda_1 & 0 \\ 0 & \lambda_2 \end{matrix} \right\| ; \qquad B : \left\| \begin{matrix} \lambda & 0 \\ 1 & \lambda \end{matrix} \right\|$$

while if the roots are complex a complex transformation will reduce the matrix to the form

$$C : \left\| \begin{matrix} \lambda & 0 \\ 0 & \bar{\lambda} \end{matrix} \right\|,$$

with a certain subcase. All told there are five cases which we shall now discuss separately.

(2.2) <u>First case.</u> <u>Real roots of same sign, matrix of type</u> A. The reduced form is

$$\frac{du}{dt} = \lambda_1 u, \qquad \frac{dv}{dt} = \lambda_2 v. \tag{2.2.1}$$

Suppose first λ_1, λ_2 both <u>negative</u>: $\lambda_i = -\mu_i$, $\mu_i > 0$. Then the general solution is

$$u = \alpha e^{-\mu_1 t}, \qquad v = \beta e^{-\mu_2 t}$$

where α, β are arbitrary constants. Evidently the characteristic γ tends to the origin as $t \to +\infty$. It reduces to the u axis when $\beta = 0$, to the v axis when $\alpha = 0$. Supposing $\alpha\beta \neq 0$ and $\mu_1 < \mu_2$ the ratio $\frac{v}{u} \to 0$ as $t \to +\infty$. Hence γ is tangent to the u axis at the origin. The form of the characteristics is that shown in fig. 2. The arrows indicate the direction of motion on the characteristics. The critical point thus arising is called a <u>stable node</u>.

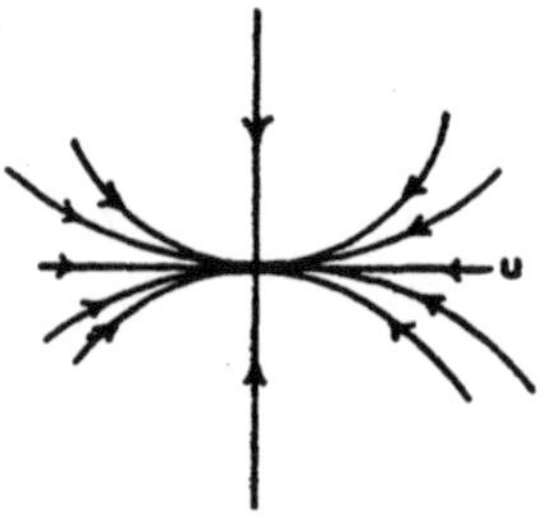

Fig. 2 Stable Node

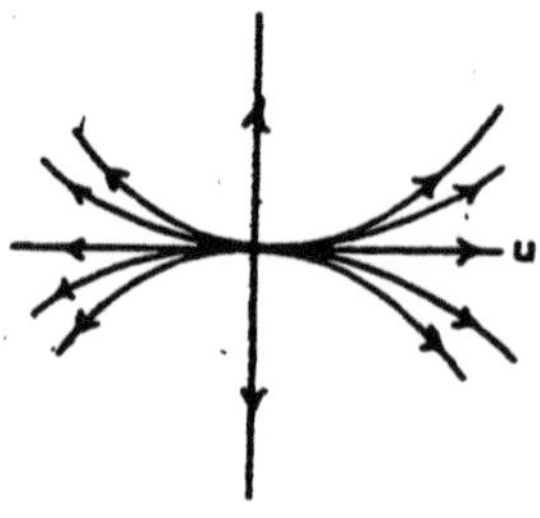

Fig. 3 Unstable Node

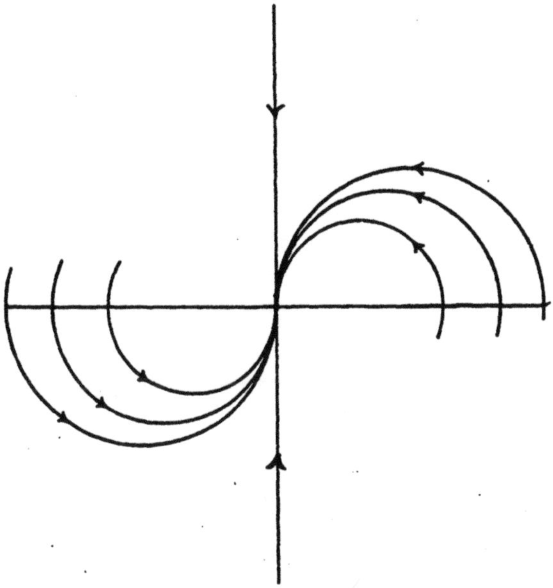

Fig. 4 Stable Node

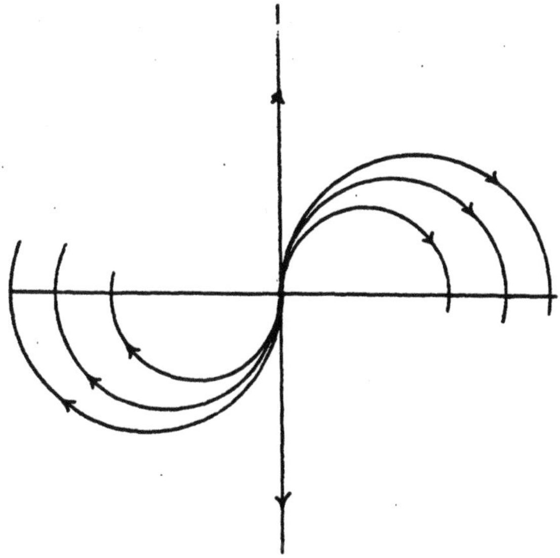

Fig. 5 Unstable Node.

If λ_1, λ_2 are real and positive, the preceding behavior corresponds to $t \to -\infty$. Hence assuming again $\lambda_1 < \lambda_2$ we have the situation of fig. 3 and the critical point known as <u>unstable node</u>.

When $\mu_1 > \mu_2$ the role of the two axes is reversed but the general aspect of the characteristics is the same. When $\lambda_1 = \lambda_2$ hence $\mu_1 = \mu_2$ all the characteristics are straight lines through the origin.

(2.3) <u>Second case, matrix of type</u> B. There is only one characteristic root λ and it is of course real. The reduced form is

$$\frac{du}{dt} = \lambda u, \qquad \frac{dv}{dt} = u + \lambda v. \tag{2.3.1}$$

Suppose first $-\lambda = \mu > 0$. The general solution is

$$u = \alpha e^{-\mu t}, \qquad v = (\alpha t + \beta)e^{-\mu t}.$$

For $\alpha = 0$ it represents the v axis. Whatever α, β both $u, v \to 0$ as $t \to +\infty$. Hence the characteristic γ tends to the origin as $t \to +\infty$. Assume now $\alpha \neq 0$. Since $\frac{v}{u} \to \infty$ when $t \to +\infty$, γ is tangent to the v axis at the origin. It also crosses the u axis at $t = \frac{-\alpha}{\beta}$. The coordinate v has an extremum when $\frac{dv}{dt} = 0$ or $t = \frac{\alpha - \beta\mu}{\alpha\mu}$. Hence the characteristics behave as indicated in fig. 4. When $\lambda > 0$ the situation is that of fig. 5, the arrows being merely reversed.

The critical point is still called a stable or unstable node. Thus the earmark of the node is a set of characteristics tending to the origin when $t \to +\infty$ for the stable node or $t \to -\infty$ for the unstable node.

(2.4) <u>Third case. Roots real and of opposite signs.</u> The matrix is then necessarily of type A. The reduced form is still (2.1.1). Assuming $\lambda_1 = -\lambda$, $\lambda_2 = \mu$, λ and $\mu > 0$, the characteristics are

$$\gamma : u = \alpha e^{-\lambda t}, \ v = \beta e^{\mu t}.$$

The axes u, v are still the characteristics corresponding to $\beta = 0$, $\alpha = 0$, If $\alpha\beta \neq 0$ then $u \to 0$, $v \to \infty$, as $t \to +\infty$. Hence the characteristics have the general form of fig. 6. The origin is then called a <u>saddle point</u>.

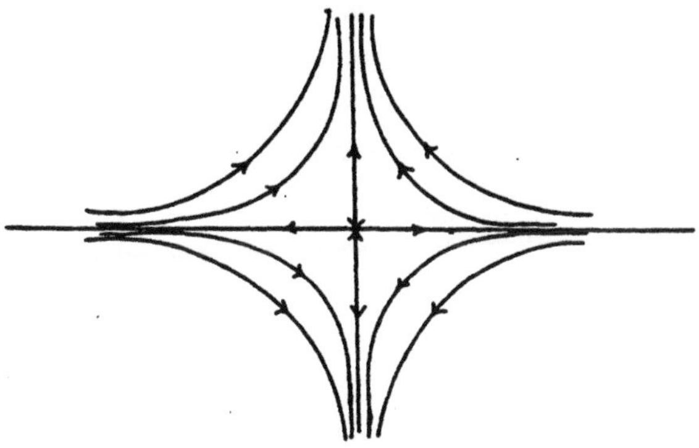

Fig. 6 Saddle point

When the signs of λ_1, λ_2 are reversed the role of the axes is reversed, or equivalently all the arrows in fig. 6 are reversed. The essential aspect of the critical point remains however the same.

(2.5) <u>Fourth case</u>: <u>Complex roots with non-zero real parts</u>. The reduction is to the form

(2.5.1) $$\frac{du}{dt} = \lambda u, \qquad \frac{d\bar{u}}{dt} = \overline{\lambda u}.$$

The transformation of coordinates

$$x \to \frac{u+\bar{u}}{2}, \qquad y \to \frac{u-\bar{u}}{2i}$$

is real and the real points of the initial system correspond to $\bar{u}$ conjugate of u. This is then assumed throughout.

Let us suppose first $\lambda = -\mu + i\omega$, where ω and μ are positive. The general solution of (2.5.1) is

$$u = \gamma e^{\lambda t} = \alpha e^{(-\mu+i\omega)t+i\beta}, \quad \alpha \text{ and } \beta \text{ real}.$$

Setting $u = re^{i\theta}$ we have then

$$r = \alpha e^{-\mu t}, \qquad \theta = \beta + \omega t$$

which represents a logarithmic spiral. Thus the aspect of the characteristics is that of fig. 7. The critical point is then called a stable focus. When $\lambda = \mu + i\omega$, $\mu > 0$, the situation is the same with arrows reversed (fig. 8) and we have the unstable focus.

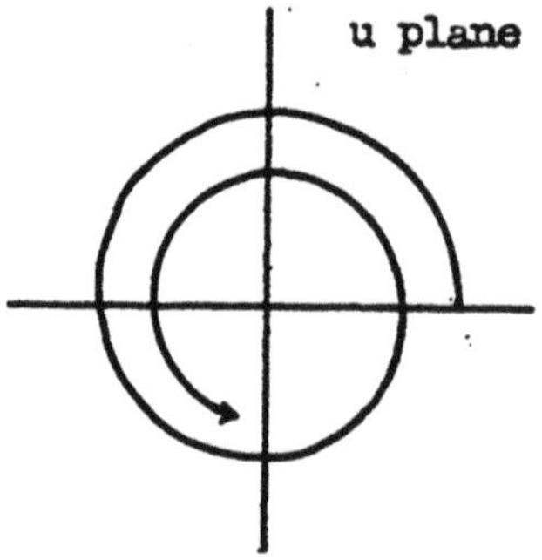

Fig. 7 Stable Focus

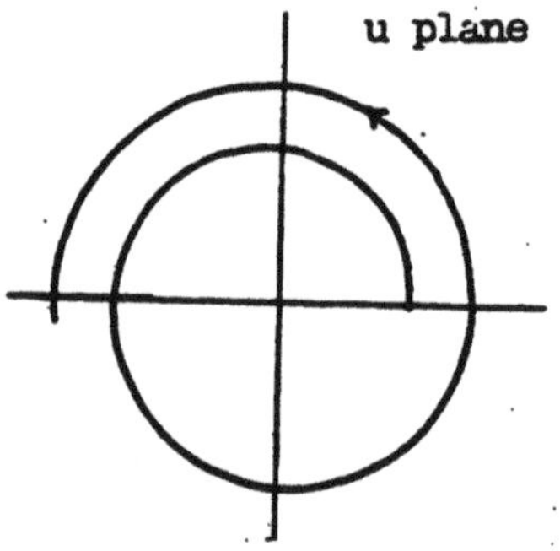

Fig. 8 Unstable Focus

(2.6) Fifth case: Pure complex characteristic roots. The situation is the same save that $\mu = 0$. Hence the characteristics are given by $r = \alpha$, $\theta = \beta + \omega t$. In other words in the (u,v) plane they are circles with the critical point as center and all described at the same angular velocity ω. The critical point is then known as a center.

(2.7) It should be kept in mind that the terms logarithmic spirals, circles, applied to the characteristics in the u plane are only partly appropriate. For even if the original x,y coordinates had been chosen rectangular, the transformations utilized were not orthogonal but merely linear homogeneous. Thus the circles in the u plane of fig. 9 would merely correspond in the original x,y plane to a family of concentric similar ellipses.

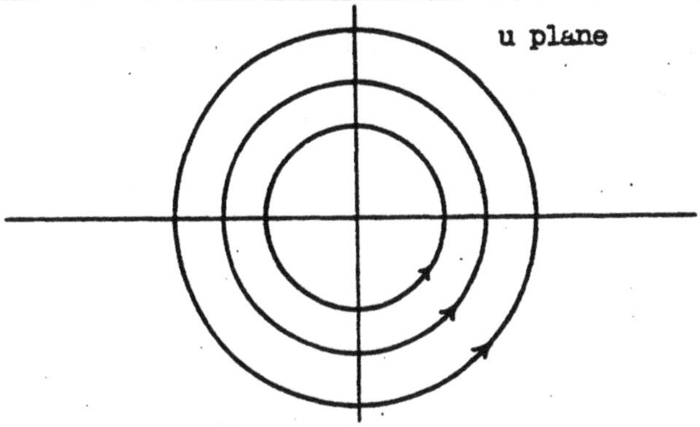

Fig. 9 Center

§3. CRITICAL POINTS IN THE GENERAL CASE

3. We shall assume now that we have the general system (1.1) with $ad - bc \neq 0$. A linear transformation of variables will reduce it to a system whose first approximation is of one of the forms considered in (2). The same terminology: node, focus, etc., is used as before. The basic result that will be proved is:

(3.1) Theorem. The behavior of the characteristics in the neighborhood of a critical point is the same as for the first approximation except that when the characteristic roots are pure complex there may arise a center or a focus.

(3.2) The nature of the coefficient matrix of (2.1) and of the roots λ_1, λ_2 gives rise to the same classification into five cases as before and we must again examine each of these cases.

4. (4.1) First case. Let the reduced system be written

$$(4.1.1)\quad \frac{dx}{dt} = \lambda_1 x + p_2(x,y), \qquad \frac{dy}{dt} = \lambda_2 y + q_2(x,y)$$

where λ_1, λ_2 are real and of the same sign. The first approximation is (2.1). According to Liapounoff's theorem (IV, 10.1) if λ_1, λ_2 are negative and (u,v) is the general solution of (2.1) then in the neighborhood of the origin the general solution of (4.1.1) is represented by

(4.1.2) $x = u + r_2(u,v), \qquad y = v + s_2(u,v).$

Since this transformation is regular at the origin the behavior of the latter as a critical point for (4.1.1) is the same as for (2.2.1) or we have a stable node. If λ_1, λ_2 are both positive we replace t by $t' = -t$, hence λ_i by $-\lambda_i$. Thus with t' as the new time we have a stable node. Hence relative to t itself the critical point is an unstable node. In both cases then the behavior relative to the critical point for the given system is the same as for its first approximation.

(4.2) Second Case. The basic system is now

(4.2.1) $\frac{dx}{dt} = \lambda x + p_2(x,y), \quad \frac{dy}{dt} = x + \lambda y + q_2(x,y)$

and the first approximation is (2.3.1). Suppose first $-\lambda = \mu > 0$ and let $u = ae^{-\mu t}$, $v = (at + b)e^{-\mu t}$ be the general solution of (2.3.1). Then according to Theorem (IV, 11.2) the general solution of (4.2.1) is given in the neighborhood of the origin by expressions

$$x = \sum_{m=1}^{+\infty} P_m(a,b;t)e^{-m\mu t}$$

$$y = \sum_{m=1}^{+\infty} Q_m(a,b;t)e^{-m\mu t}$$

where P_m, Q_m are polynomials in t whose coefficients are forms of degree m in a,b. The first terms of the x,y series are respectively equal to u,v. Hence we may write

$$x = ae^{-\mu t} + \sum_{m>1} P_m e^{-m\mu t}$$

$$y = (at+b)e^{-\mu t} + \sum_{m>1} Q_m e^{-m\mu t}.$$

This shows that (x,y) tends to the origin as $t \to +\infty$ and that as $t \to +\infty$

$$\lim \frac{y}{x} = \lim \frac{at+b}{a} = \infty .$$

From this follows readily that the behavior of the characteristics is the same as for the first approximation and is represented near the origin by fig. 4 when $\lambda < 0$.

If $\lambda > 0$ we replace again t by $t' = -t$, and as in the preceding case, show that the behavior of the characteristics is the same as in fig. 3 with arrows reversed i.e., the same as in fig. 5. Thus we have the same stable or unstable node as for the first approximation.

(4.3) Third Case. The basic system is

$$(4.3.1)\quad \frac{dx}{dt} = -\lambda x + p_2(x,y), \quad \frac{dy}{dt} = \mu y + q_2(x,y), \quad \lambda, \mu > 0$$

and the first approximation is (2.4.1). The latter has the special solutions

$$(4.3.2)\qquad u = e^{-\lambda t}, \qquad v = 0$$

$$(4.3.3)\qquad u = 0, \qquad v = e^{\mu t}.$$

Referring to (IV, 10.9, 10.11), there corresponds to (4.3.2) a solution of (4.3.1) whose form for t sufficiently large (i.e., as $t \to +\infty$) is

$$x = u + f_2(u), \qquad y = g_2(u), \qquad u = e^{-\lambda t}$$

and another whose form for t sufficiently small (i.e., as $t \to -\infty$) is

$$x = h_2(v), \qquad y = v + k_2(v), \qquad v = e^{\mu t}.$$

The transformation of variables

$$(4.3.4)\qquad x = u + f_2(u), \qquad y = v + k_2(v)$$

is analytic and regular at the origin. Hence it will not modify the behavior of the characteristics in a sufficiently small neighborhood of the origin. The inverse of (4.3.4) is of the same form

$$(4.3.5)\qquad u = x + r_2(x), \qquad v = y + s_2(y)$$

and behaves in similar manner (see 1.5). Passing to the variables u,v the initial system (4.3.1) is replaced by a new system

$$(4.3.6)\ \frac{du}{dt} = -\lambda u + F_2(u,v), \qquad \frac{dv}{dt} = \mu v + G_2(u,v)$$

satisfied identically by $u = 0$, $v = e^{\mu t}$. This implies that $F_2(0,v) = 0$ for $v = e^{\mu t}$, i.e. for v real and arbitrary positive. Since $F_2(0,v)$ is analytic in v and has an arc of zeros it is identically zero. Hence $F_2(u,v) = uF_1(u,v)$. Similarly since the system is satisfied for $u = e^{-\lambda t}$, $v = 0$, $G_2(u,v) = vG_1(u,v)$. Thus the system (4.3.6) is of the form

$$(4.3.7)\ \frac{du}{dt} = u(-\lambda+F_1(u,v)), \qquad \frac{dv}{dt} = v(\mu+G_1(u,v)).$$

Referring again to (1.5) this last system may take the place of the initial system. Notice that now the special solutions are $u = 0$, $v = 0$.

Since for the new system the origin is still an isolated critical point we may assume ρ so small that there are no other critical points than the origin interior to the circle $C : u^2 + v^2 = \rho^2$. As a consequence in the vicinity of any point $(u_0,0)$, $0 < u_0 < \rho$, the characteristics have the general behavior indicated in fig. 1, where A is now the point $(u_0,0)$ and γ is the u axis. Hence $v = 0$ is the only characteristic $\to (u_0,0)$. Thus within the circle C the characteristics other than the axes have no limit-point on an axis.

Since F_1, G_1 are continuous in the vicinity of the origin and $\to 0$ when (u,v) tends to the origin, we may assume ρ so small that F_1, G_1 are continuous and $|F_1| < \lambda$, $|G_1| < \mu$ within C. As a consequence along any characteristic γ within C, $\frac{du}{dt}$ has the sign of $-u$ and $\frac{dv}{dt}$ has the sign of v. Since γ within C cannot tend to the axes it has the general behavior indicated in fig. 6 i.e. we have a saddle point as for the first approximation.

In the literature there will be found proofs of property

(4.3.8) <u>When the roots λ_1, λ_2 are real and of opposite signs there are exactly two characteristics tending to the origin and they are not tangent</u>.

A glance at fig. 5 shows that thie property is true.

5. <u>Fourth and fifth cases</u>. These last two cases may be conveniently taken up together. The fundamental system is now

$$(5.1) \qquad \frac{dx}{dt} = \lambda x + p_2(x,\bar{x}), \quad \frac{d\bar{x}}{dt} = \overline{\lambda x} + \bar{p}_2(\bar{x},x)$$

and the first approximation is (2.5.1). Here of course the series may have complex coefficients and $\bar{p}_2$ indicates that each of these has been replaced by its conjugate. For the real points x and $\bar{x}$ are conjugate.

The most convenient method for dealing with the characteristics is to pass to polar coordinates r, θ. Write explicitly $\lambda = \mu + i\omega$. Then (5.1) yields

$$e^{i\theta}\frac{dr}{dt} + ire^{i\theta}\frac{d\theta}{dt} = (u + i\omega)\, re^{i\theta} + p_2(re^{i\theta}).$$

Upon dividing by $e^{i\theta}$ and equating real and complex parts we obtain relations

$$(5.2) \qquad \frac{dr}{dt} = \mu r + \alpha_1(\theta)r + \alpha_2(\theta)r^2 + \ldots$$

$$(5.3) \qquad \frac{d\theta}{dt} = \omega + \beta_1(\theta)r + \beta_2(\theta)r^2 + \ldots$$

where $\alpha_n(\theta)$, $\beta_n(\theta)$ are real forms of degree n in cos θ, sin θ. The series converge for r sufficiently small and any θ. Since $\omega \neq 0$, we obtain by division a relation

$$\frac{dr}{d\theta} = r\left(\frac{\mu}{\omega} + \gamma_1(\theta)r + \dots\right) \tag{5.4}$$

where the coefficients and convergence are as before. Since the system is analytic the solution, taking the value ρ for $\theta = \theta_0$, may be represented in the form (see II, 9.1)

$$r(\theta,\theta_0,\rho) = c_1(\theta,\theta_0)\rho + c_2(\theta,\theta_0)\rho^2 + \dots$$

the series being valid for an arbitrary θ range and ρ small. The term independent of ρ is missing since $\rho = 0$ must yield the solution $r = 0$. Since the series may be assumed valid for $\theta = 0$, the solutions near the origin may be assumed to have their initial value ρ for $\theta_0 = 0$. Hence we may write our solutions $r(\theta,\rho)$ and choose for them a representation

$$r(\theta,\rho) = c_1(\theta)\rho + c_2(\theta)\rho^2 + \dots . \tag{5.5}$$

The c_n are determined by substituting in (5.4) and identifying equal powers of ρ. We thus obtain a system

$$\begin{cases} \dfrac{dc_1}{d\theta} = \dfrac{\mu}{\omega}c_1 \\ \dfrac{dc_2}{d\theta} = \dfrac{\mu}{\omega}c_2 + \gamma_2(\theta)c_1^2 \\ \dots\dots\dots\dots \end{cases} \tag{5.6}$$

Since $r(0,\rho) = \rho$ identically we must have

$$c_1(0) = 1, \qquad c_n(0) = 0 \text{ for } n > 1. \tag{5.7}$$

The differential equations (5.6) together with the initial

conditions (5.7) enable one to determine the $c_n(\theta)$ one at a time. In particular

$$(5.8) \qquad c_1(\theta) = e^{\frac{\mu}{\omega}\theta}$$

In order that $r(\theta,\rho)$ be periodic of period 2π, or which is the same in order that its characteristic be an oval surrounding the origin we must have $r(2\pi,\rho) = \rho$ or

$$(5.9) \qquad \rho = \sum c_n(2\pi)\rho^n = \varphi(\rho).$$

There are now two possibilities:

(5.10) <u>All the characteristics for ρ sufficiently small are ovals surrounding the origin</u>, i.e. <u>the origin is a center</u>. Then (5.9) is satisfied identically. Since $\varphi(\rho) - \rho$ is holomorphic at $\rho = 0$, all its coefficients vanish. Hence

$$c_1(2\pi) = 1, \qquad c_n(2\pi) = 0 \text{ for } n > 1.$$

The first relation yields $e^{\frac{2\pi\mu}{\omega}} = 1$, hence $\mu = 0$ and the rest yield the condition that every $c_n(\theta)$ has the period 2π. Thus the center can only arise whenever the characteristic roots are pure complex and furthermore the $c_n(\theta)$ are all periodic. Conversely when these two conditions hold $\varphi(\rho) - \rho = 0$, the characteristics near the origin are ovals surrounding it and the origin is a center.

Notice that when the origin is a center the time period T for the description of the characteristic $r(\theta,\rho)$ calculated by means of (5.3) is

$$T = \int_0^{2\pi} \frac{d\theta}{\omega + \beta_1(\theta)r(\theta,\rho) + \dots}$$

and will generally depend upon ρ. Thus it is not necessarily constant, contrary to what happens in the linear case. Expressed also in another way the point $r(t)$, $\theta(t)$ of a given characteristic γ describes it with an instantaneous or even an average angular velocity which are not generally independent of γ.

(5.11) The coefficients $c_n(\theta)$ do not all have the period 2π. The closed characteristics, if any exist in the range considered, will correspond to the solutions in ρ of (5.9). Under our hypothesis the coefficients of $\varphi(\rho) - \rho$ are not all zero. Hence $\rho = 0$ is an isolated root of (5.9). This means that a $\sigma > 0$ may be selected such that (5.9) has no roots in the interval $0 < \rho < \sigma$. Let ρ be confined henceforth to this interval.

Starting at the point $P(0,\rho)$ of a characteristic, as θ varies by 2π, 4π, ..., there will be reached points $P_1(0,\rho_1)$, $P_2(0,\rho_2)$, ... Suppose $\rho_1 < \rho$. Then necessarily $\rho > \rho_1 > \rho_2$ ... . Thus $\{\rho_n\}$ has a limit η, $0 \leq \eta < \sigma$. Let us suppose $\eta > 0$. Since $\rho_{n+1} = \varphi(\rho_n)$, the sequence $\{(\varphi(\rho_n)-\rho_n)\} \to 0$. Since $\varphi(\rho) - \rho$ is continuous at $\rho = \eta$, we have $\varphi(\eta) - \eta = 0$ or (5.9) has a root in the interval $0 < \rho < \sigma$. Since this contradicts the assumption on σ we must have $\eta = 0$. Hence the characteristics behave as in fig. 7. The critical point is a stable focus. If $\rho_1 > \rho$ the situation is analogous save that we merely conclude that the spirals diverge from the origin. The latter is then an unstable focus.

Suppose in particular $\mu \neq 0$. If $\mu < 0$ then for r small $\frac{dr}{d\theta}$ will steadily decrease and we have the stable focus, while for $\mu > 0$ it will be the reverse with an unstable focus. Thus in the fourth case, - real parts of the characteristic roots non zero - the situation is the same as for the first approximation.

This completes the proof of (3.1).

§4. THE INDEX IN THE PLANE

6. (6.1) Returning to the basic system (1.1) we shall denote by $\vec{V}(M)$ the vector in the plane Π of x,y whose initial point is M(x,y) and whose components are P(x,y), Q(x.y). Thus $\vec{V}(M) = 0$ when and only when M is a critical point. The vector distribution over the plane Π, thus arising, makes up a certain field $\mathcal{F}$. It will be convenient to confine our attention mainly to this field $\mathcal{F}$ and to utilize wherever possible its analyticity (its components are analytical). However a good part of the arguments and all of the results are valid for very general fields.

(6.2) We shall have occasion to consider various Jordan curves (topological images of the circumference) in the plane Π. It will be understood once and for all that all Jordan curves considered are "piecewise" analytic, i.e., made up of a finite number of consecutive closed analytic arcs A_1B_1, B_1C_1, ..., D_1A_1 related like the sides of a simple polygon (fig. 10). Any one of the arcs for example A_1B_1 has a parametric representation of

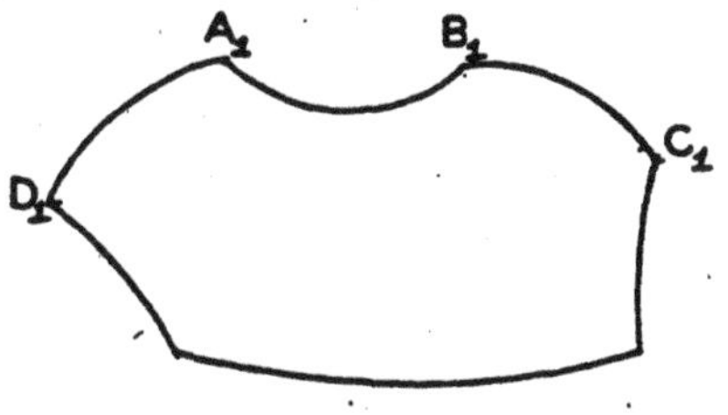

Figure 10

the form

$$x = f(u), \qquad y = \varphi(u), \qquad 0 \leq u \leq 1,$$

where f and φ are analytic on their range and their derivatives never vanish simultaneously. On each arc then there is a tangent and it turns continuously when the corresponding parameter u varies on its range. If the curve J is analytic then the tangent turns continuously as its point of contact describes J. A closed characteristic is an example of an analytic Jordan curve.

(6.3) Let the plane Π be oriented by assigning a positive direction of rotation for its angles. This assigns a positive sense of description or orientation to its convex ovals. We shall also admit that it assigns likewise a positive orientation to the Jordan curves here considered. This property, as well as the standard Jordan curve theorem (separation of the plane into two regions) are easily proved for our special Jordan curves.

(6.4) A Jordan curve which does not pass through the critical points will be referred to as an *allowable* Jordan curve.

7. *The index of Jordan curves*. (7.1) Let then J be an allowable Jordan curve and let a point M describe J once positively. The vector $\vec{V}(M)$ is defined at each position of M on J. As M, starting from any point M_0 of J describes J once positively, the angle θ which the vector makes with a fixed direction varies by a multiple

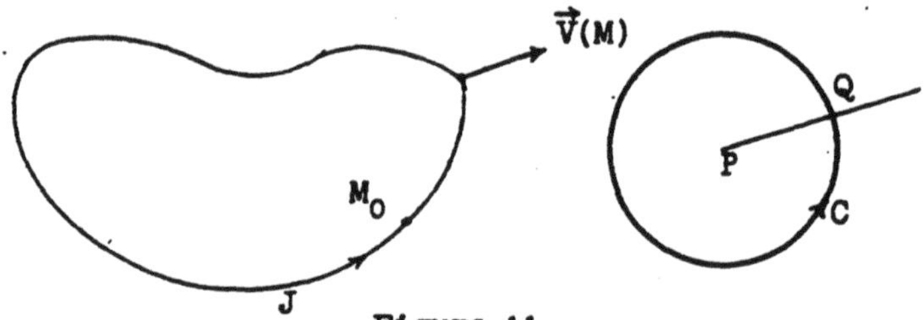

Figure 11

$2m\pi$ of 2π, where m depends solely upon the field $\mathcal{F}$ and the curve J. In fact when the orientation of Π, hence also that of J, is reversed, m still remains the same. The number m is called the *index of* J *relative* to $\mathcal{F}$, or merely the *index* of J, written Index (J).

(7.2) The index is also conveniently expressed as an integral:

$$(7.3)\quad \text{Index }(J) = \oint_J d \arctan \frac{Q}{P} = \oint_J \frac{PdQ - QdP}{P^2 + Q^2}.$$

Since the integrand is a total differential and continuous except at the critical points we have:

(7.4) <u>The index of J does not change when J varies continuously without crossing critical points</u>.

(7.5) An equivalent but more geometric way of defining the index is as follows: -- Take a fixed positively oriented circumference C and from its center P draw the ray parallel to the vector $\vec{V}(M)$. Let the ray intersect C at Q. Then as M describes J once, Q will describe C an algebraic number m of times and m = Index (J). It is clear that in this definition C may be replaced by any convex oval surrounding P. If the plane is subjected to an affine transformation:

$$x = ax' + by' + \alpha, \quad y = cx' + dy' + \beta,$$
$$ad - bc \neq 0,$$

the circle C is replaced by an ellipse surrounding P. Hence by the remark just made:

(7.6) <u>An affine transformation does not change the index of J</u>.

A convenient property is the following. Along J replace $\vec{V}(M)$ by a new vector $\vec{V}'(M)$ always $\neq 0$ and varying continuously with M. One may evidently define a corresponding Index (J)'. Then

(7.7) <u>If the vectors $\vec{V}(M)$, $\vec{V}'(M)$ are never in opposition on J then</u> Index (J)' = Index (J).

Take any number k such that $0 \leq k \leq 1$ and for each k define $\vec{V}''(M) = (1-k)\,\vec{V}(M) + k\vec{V}'(M)$. Since $\vec{V}'' \neq 0$ whatever k and whatever M on J, the corresponding Index (J)" may always be defined and it varies continuously with k. Since it is an integer it is constant. Hence its extreme values Index (J), Index (J)' are equal.

On the other hand we manifestly have:

(7.8) <u>If the vectors $\vec{V}(M)$, $\vec{V}'(M)$ are always in opposition along J then again</u> Index (J) = Index (J)'.

(7.9) <u>Index of critical points</u>. Let J be a small convex curve surrounding a single critical point A. By (7.4) the index of J is the same for all curves similar to J. Thus its value depends solely upon A and it is known as the index of A, written Index (A). By mere paraphrase of reasonings familiar in the theory of residues, we find:

(7.10) <u>The index of a non-critical point is zero</u>.

(7.11) <u>If J surrounds the critical points</u> $A_1, \ldots, A_r$ <u>then</u>

$$\text{Index}\ (J) = \sum \text{Index}\ (A_i).$$

This property reduces the calculation of the Index (J) to the calculation of the indices of the critical points.

8. <u>Index of the elementary critical points</u>. Let us suppose as usual that the critical point is at the origin and that the index is given by the integral (7.3) where J is a circle of small radius r. Our task will be greatly simplified as a consequence of

(8.1) <u>The index of the origin relative to the system</u> (2.1) <u>is the same as the index relative to the first approximation</u>.

The components P,Q of the vector $\vec{V}(M)$ are

$$P = ax + by + P_2(x,y), \qquad Q = cx + dy + Q_2(x,y),$$
$$\text{(8.1.1)} \qquad ad - bc \neq 0$$

and the corresponding components of the vector $\vec{V}'(M)$ for the first approximation are

$$\text{(8.1.2)} \qquad P^* = ax + by, \qquad Q^* = cx + dy.$$

According to (7.7) we merely have to show that on J, or preferably for x,y sufficiently small we cannot have a relation $k\vec{V}(M) + \vec{V}'(M) = 0$, $k > 0$, or we cannot have

$$kP + P^* = 0, \qquad kQ + Q^* = 0, \qquad k > 0,$$

or explicitly that one must rule out relations

$$(8.1.3) \quad (1+k)(ax+by) + P_2 = 0, \qquad (1+k)(cx+dy) + Q_2 = 0.$$

As a consequence of (8.1.3) we find

$$(1+k)^2[(ax+by)^2 + (cx+dy)^2] = P_4(x,y).$$

Introducing polar coordinates r, θ we obtain

$$(1+k)^2[(a \cos \theta + b \sin \theta)^2 + (c \cos \theta + d \sin \theta)^2] =$$
$$(8.1.4) \qquad = r^2\alpha_4(\theta) + r^3\alpha_5(\theta) + \dots ,$$

where $\alpha_n(\theta)$ is a form of degree n in $\sin \theta$, $\cos \theta$ and the series at the right converges for r small and any θ, also uniformly in θ. As a consequence the right hand side of (8.1.4) $\to 0$ with r whatever θ. On the other hand since $ad - bc \neq 0$, we can only have $ax + by = 0$, $cx + dy = 0$ if $x = y = 0$. Hence the square bracket at the left $\neq 0$ whatever θ, i.e., on $0 \leq \theta \leq 2\pi$. Thus it is continuous and positive on a closed interval and hence it has a positive lower bound ξ. Since $k > 0$, the left hand side $\geq \xi$ whatever θ. Hence for r sufficiently small the two sides of (8.1.4) are different. This proves (8.1).

9. We shall now deal directly with the first approximation. According to (7.6) we may assume it in reduced form. Let the reduced forms be those of (3) save that we still use coordinates x,y instead of u,v. Here again each case requires special examination.

(9.1) Case I. (Node). Here λ_1, λ_2 are real and of the same sign. Assuming $|\lambda_2| \geq |\lambda_1|$ we have on C: $x^2 + y^2 = r^2$, $P = \lambda_1 x$, $Q = \lambda_2 y$, $P^2 + Q^2 = \lambda_1^2 x^2 + \lambda_2^2 y^2$. It follows that if we vary λ_1, λ_2 continuously, without vanishing, till they both become +1 when they are positive, -1 when they are negative, the integral (7.3) will vary continuously also. Since it is an integer it is then constant and so the modifications of λ_1, λ_2 do not change Index (A). At the end the vector $\vec{V}(M)$ will be along the radius AM and always pointing inwards or always outwards. In both cases the index is unity.

(9.2) Case II. (Node). Here $\lambda_1 = \lambda_2 = \lambda$. Instead of the circle C we shall take the curve J of (7.3) as indicated in fig. 12: J = BCDEB'C'D'E'B'. Here CDE, C'D'E' are portions of characteristics and the other parts of J are segments. The figure is symmetric with respect to the origin and it is drawn for $\lambda > 0$ (unstable node). It is clear that J is deformable continuously into a circumference and so it is admissible as

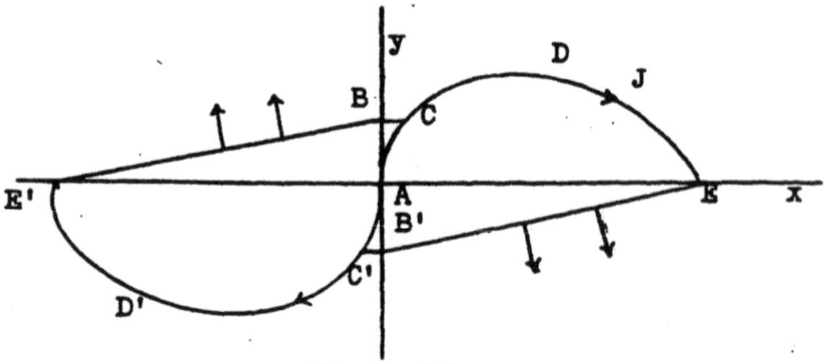

Figure 12

curve J in (7.3). The angular variation of $\vec{V}(M)$ along B'EDCB is manifestly π, since the vectors turn continuously forward from $\frac{-\pi}{2}$ to $\frac{+\pi}{2}$. Similarly on the second part of J from B to B'. Hence the total angular variation of $\vec{V}(M)$ is 2π and the index is again unity. If $\lambda < 0$ all the vectors are reversed and the result is the same.

(9.3) Case III. (Saddle point). This time λ_1, λ_2 are real but of opposite signs. We may assume $\lambda_1 < 0$, $\lambda_2 > 0$ and choose an integration curve J which is the arc BCDEF of fig. 13 repeated by symmetry around the axes. Along that arc the angular variation of $\vec{V}(M)$ is $\frac{-\pi}{2}$, hence its total angular variation is -2π. Hence the index of the saddle point is -1.

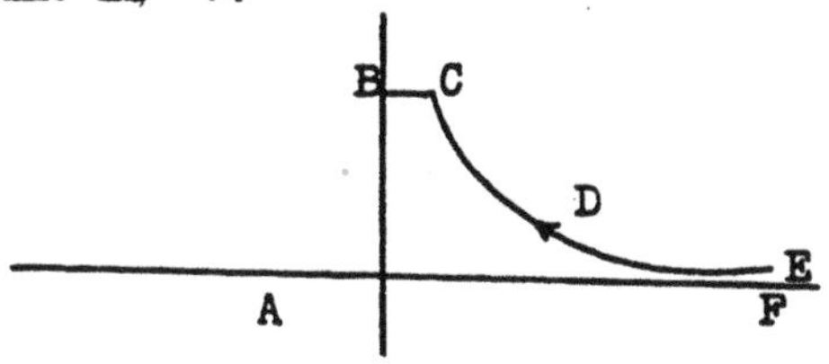

Figure 13

(9.4) Cases IV and V (focus and center). Let the reduced form be this time

$$\frac{dx}{dt} = (\mu + i\omega)x, \qquad \frac{d\bar{x}}{dt} = (\mu - i\omega)\bar{x}.$$

The denominator in the integral (7.3) is the squared length of the vector $\vec{V}(M)$. Its value here for M on the circle of radius one is $\mu^2 + \omega^2$. Hence the same continuity argument as before shows that we may vary μ continuously to zero, and ω continuously to unity without changing Index (A). For $\omega = 1$ the vector $\vec{V}(M)$ is represented in the complex plane by ix, i. e., it is tangent to the circle. As M describes the latter $\vec{V}(M)$ rotates forward and so the angular variation is 2π. Thus the index of a focus or a center is unity.

To sum up we have proved:

(9.5) Theorem. The index of an elementary critical point other than a saddle point is unity; the index of a saddle point is (-1).

10. One will readily surmise property

(10.1) The index of a Jordan curve J is invariant under a transformation of coordinates which is regular in a 2-cell containing J.

In view of (7.10) this will follow immediately from

(10.2) The index of a point is unchanged by a transformation regular at the point.

We may assume as usual that the point in question is the origin and that the regular transformation is

$$(10.3)\qquad \begin{aligned} x' &= \alpha x + \beta y + p_2(x,y) \\ y' &= \gamma x + \delta y + q_2(x,y) \\ &\alpha\delta - \beta\gamma \neq 0. \end{aligned}$$

Under our assumption the matrix $A = \begin{Vmatrix} \alpha & \beta \\ \gamma & \delta \end{Vmatrix}$ has an inverse $A^{-1} = \begin{Vmatrix} \alpha_1 & \beta_1 \\ \gamma_1 & \delta_1 \end{Vmatrix}$. Consider now the linear transformation

$$(10.4)\qquad x'' = \alpha_1 x' + \beta_1 y', \qquad y'' = \gamma_1 x' + \delta_1 y'.$$

Since the transformation (10.4) does not affect indices we may replace (10.3) by its product by (10.4) at the left. Replacing x'', y'' by x', y' the new transformation will have the form

$$(10.5)\qquad x' = x + p_2(x,y), \qquad y' = y + q_2(x,y).$$

If this is applied to (1.1) there results a similar system with P,Q replaced by

$$(10.6)\qquad \begin{aligned} P'(x',y') &= \left(1 + \frac{\partial p_2}{\partial x}\right) P + \frac{\partial p_2}{\partial y} Q = (1+\lambda)P + \mu Q \\ Q'(x',y') &= \frac{\partial q_2}{\partial x} P + \left(1 + \frac{\partial q_2}{\partial y}\right) Q = \nu P + (1+\rho)Q \end{aligned}$$

where λ, μ, ν, ρ have their obvious meaning. All that one needs to know concerning them is that they are continuous and $\rightarrow 0$ as the point M tends to the origin.

Let $\vec{V}(M)$, $\vec{V}'(M)$ denote the vectors whose respective components are (P,Q), (P',Q'). To prove (10.2) we have to show that upon replacing $\vec{V}$ by $\vec{V}'$ the index of the origin does not change. As in (8.1) this will follow if we can show that when M remains in a suitably small neighborhood U of the origin O, and M is not the origin then we cannot have a relation $k\vec{V}(M) + \vec{V}'(M) = 0$, or equivalently that under the same conditions

$$(1+k+\lambda)P + \mu Q = 0, \qquad \nu P + (1+k+\rho)Q = 0$$

cannot hold. Since the origin is an isolated critical point when U is sufficiently small and M is as stated, P and Q are not both zero. Hence we must have

$$(10.7) \qquad (1+k+\lambda)(1+k+\rho) - \mu\nu = 0, \qquad k > 0.$$

Now for M at the origin $\lambda = \mu = \nu = \rho = 0$ and (10.7) has the only solution $k = -1$ in that variable. Hence for $M \in U$ and U sufficiently small the solutions of (10.7) in k are very near (−1) and hence negative. In other words (10.7) cannot be satisfied under the conditions considered. This proves (10.2) and hence (10.1).

(10.8) Application: -- Theorem. The index of a closed characteristic is unity.

Immediate corollaries are:

(10.9) A closed characteristic must surround some critical point.

(10.10) If a closed characteristic surrounds only elementary critical points then they cannot all be saddle points.

(10.11) Proof of (10.8). Let γ be a closed characteristic and Ω its interior. Since γ is analytical it is a classical theorem that there exists a conformal mapping φ of $\bar{\Omega} = \gamma \cup \Omega$ onto a circumference C and its interior U and φ is regular on $\bar{\Omega}$. Hence Index (γ) = Index (C). But along C the vector $\vec{V}(M)$ is tangent to C. Hence it turns continuously in the same direction and so its angular variation is 2π. Hence Index (C) = 1 = Index (γ).

(10.12) A general observation. Up to the present, it has been assumed for simplicity that P, Q in the system (1.1) are entire functions. It is obvious however that all the results obtained go through when P, Q are merely holomorphic in a certain closed circular region Ω

provided that the only critical points and arcs of characteristics considered are those in Ω.

§5. DIFFERENTIAL SYSTEMS ON A SPHERE

11. The proper study of the behavior of the characteristics throughout the plane Π can scarcely be made without a thorough knowledge of their behavior at infinity. One is therefore led to utilizing some method for closing up Π at infinity. To close Π by a line and turn it into a projective plane $\mathcal{P}$ is unsatisfactory. For if one follows a line L in the two opposite directions, the two limiting positions of the vector $\vec{V}(M)$ may be in opposition. As a consequence one may then only extend the basic field $\mathcal{J}$ to $\mathcal{P}$ by describing the latter twice. The true configuration then obtained is the doubly covering surface of $\mathcal{P}$ which is a sphere. Thus it is best to have recourse, with Poincaré, directly to the sphere.

It must be added that the extension just described is really useful only when the functions P, Q in (1.1) are polynomials. This restriction is imposed throughout the present section.

(11.1) Let the plane Π in which we have operated so far be identified with the plane $z = 1$ in an Euclidean space $\mathcal{E}^3$ referred to coordinates x, y, z. Now to a point $M(x,y,z)$ of $\mathcal{E}^3 - 0$ (0 is the origin) there corresponds the intersection $M^*(\xi,\eta,1)$ of OM with Π, where $\xi = \frac{x}{z}$, $\eta = \frac{y}{z}$. We denote by π the projection $\mathcal{E}^3 - 0 \to \Pi$ such that $\pi M = M^*$. The inverse image of M^* consists of all the points of the line OM except 0 and they are all represented by $(k\xi, k\eta, k)$, $k \neq 0$.

(11.2) If $f(x,y)$ is a polynomial of degree n, we may make it homogeneous in x,y,z by replacing any term in $x^p y^q$ by $x^p y^q z^{n-p-q}$. The resulting form of degree n will be written $f(x,y,z)$. Conversely given such a form we write $f(x,y)$ for $f(x,y,1)$.

Consider now the system (1.1) and let p,q be the degrees of P,Q and n the highest of p and q. Let us associate with (1.1) the differential equation

$$(11.2.1)\qquad \begin{vmatrix} dx, & dy, & dz \\ x, & y, & z \\ z^{n-p}P(x,y,z), & z^{n-q}Q(x,y,z), & 0 \end{vmatrix} = 0$$

or after expanding the determinant

$$(11.2.2)\quad -z^{n-q+1}Qdx + z^{n-p+1}Pdy + (z^{n-q}xQ - z^{n-p}yP)dz = 0.$$

Since $P(x,y)$ and $Q(x,y)$ have no common factor this holds also regarding $P(x,y,z)$ and $Q(x,y,z)$. Hence the coefficients of dx and dy have at most a power of z in common. Now if $p < q = n$ or $q < p = n$ no power of z divides the coefficient of dz and (11.2.1) is to be left as it is. On the other hand, if $p = q = n$ and $p_n(x,y)$, $q_n(x,y)$ are the terms of highest degree in $P(x,y)$, $Q(x,y)$, and if $xq_n = yp_n$ then z factors out of every term, but no higher power does. If that is the case the factor z is to be suppressed. In both cases (11.2.1) assumes the form

$$(11.2.3)\quad A(x,y,z)dx + B(x,y,z)dy + C(x,y,z)dz = 0$$

where A, B, C are forms of the same degree without any common factor.

(11.3) Observe explicitly that in both cases considered when $z = 1$, and hence $dz = 0$, (11.2.3) reduces to

$$(11.3.1)\qquad P(x,y)dy - Q(x,y)dx = 0$$

where

$$(11.3.2)\quad A(x,y) = -Q(x,y), \qquad B(x,y) = P(x,y).$$

In other words in the plane $\Pi : z = 1$, (11.2.3) always reduces to the initial system (1.1) with dt eliminated.

(11.4) It is worth while to underscore the geometric

meaning of the differential relation (11.2.3). An analytical arc λ

$$x = f(u), \qquad y = g(u), \qquad z = h(u)$$

is by definition a solution of (11.2.1), or equivalently of (11.2.3) whenever

$$(11.4.1) \qquad A\frac{df}{du} + B\frac{dg}{du} + C\frac{dh}{du} = 0$$

or the arc λ. Now if $k(u)$ is a function of u not vanishing on the arc λ then

$$x = kf(u), \qquad y = kg(u), \qquad z = kh(u)$$

is immediately seen to represent likewise an arc λ_1 in $\mathcal{E}^3$ and to be a solution of (11.2.3) when λ is one. Notice that the points (x,y,z) and (kx,ky,kz) are collinear with the origin.

12. We are evidently very close to an interpretation by means of planar projective geometry. The projective plane arises here in the familiar manner by identifying all the points (kx,ky,kz) of any line L through the origin. We observed already, at the beginning of the section that one must distinguish between the opposite ends of L. This might be obtained by identifying merely the points of the ray OL' corresponding to $k > 0$, and likewise those of its opposite OL'' corresponding to $k < 0$. By identifying each with its intersection with a fixed sphere we obtain a spherical representation which is more adequate in every way.

(12.1) Consider then the unit-sphere S:

$$(12.1.1) \qquad x^2 + y^2 + z^2 = 1$$

and its two hemispheres H: $z > 0$ and H': $z < 0$. Let λ be an arc in one of the two, say in H, and let λ^* be its projection $\pi\lambda$ in the plane Π. Referring to (11.3) if λ is an arc of characteristic of (11.2.3) (i.e. satisfies that relation, or equivalently (11.4.1)) then λ^* is an arc of characteristic of the basic system (1.1). Conversely the argument of (11.4) shows that if λ^* is an arc of characteristic of (1.1) then the two arcs λ, λ' in H, H' making up $\pi^{-1}\lambda^*$ are arcs of characteristic of (11.2.3). Thus it is natural to consider the differential relation (11.2.3) associated with the equation (12.1.1) of the sphere as the extension to the sphere of the system (1.1).

(12.2) We have then for the characteristics on the sphere the system of two equations

$$A \frac{dx}{dt} + B \frac{dy}{dt} + C \frac{dz}{dt} = 0 \tag{12.2.1}$$

$$x \frac{dx}{dt} + y \frac{dy}{dt} + z \frac{dz}{dt} = 0. \tag{12.2.2}$$

The critical points of the system on S are by definition the points which satisfy $A = B = C = 0$.

(12.3) <u>The arcs of the equator</u> $z = 0$ <u>between consecutive critical points on the equator are arcs of characteristics. Hence an elementary critical point on the equator can only be a node or a saddle point but not a focus or a center.</u>

For $z = 0$, $dz = 0$ satisfies (11.2.1).

(12.4) <u>The figure composed of the characteristics is symmetric relative to the origin.</u>

For the coefficients in the two equations above are homogeneous and hence the equations are unchanged upon reversing the signs of x, y, z.

(12.5) <u>The number of critical points is finite and even. Moreover they are antipodal in pairs.</u>

Since A, B, C have no common factor the cones $A = 0$, $B = 0$, $C = 0$ intersect in a finite number of straight lines through the origin and this implies (12.5).

(12.6) For convenience select as solution of the linear system (12.2.1), (12.2.2) in $\frac{dx}{dt}$, ..., the simplest solution

$$(12.6.1)\quad \frac{dx}{dt} = Bz - Cy,\quad \frac{dy}{dt} = Cx - Ay,\quad \frac{dz}{dt} = Ay - Bx$$

and let $\vec{V}(M)$ denote the vector whose components are the right-hand sides of (12.6.1). It is clear that $\vec{V}(M)$ is unique and continuous in M. Moreover, owing to (12.2.2), $\vec{V}(M)$ is in the plane tangent to the sphere at M. Thus it may be said to give rise to a vector distribution on the sphere.

(12.7) $\vec{V}(M) = 0$ <u>at the critical points and nowhere else</u>.

Let $\vec{W}(M)$ be the vector whose components are (A,B,C). The critical points have been defined as the zeros of $\vec{W}(M)$. Now in vector notations (cross-product) (12.6.1) means that $\vec{V}(M) = \vec{r} \times \vec{W}(M)$ where $\vec{r}$ is the radial vector (x,y,z). Since $\vec{r} \neq 0$, $\vec{V}(M)$ will vanish whenever: (a) $\vec{W}(M) = 0$, i.e. at the critical points, or else (b) $\vec{W}(M)$ is a non-zero radial vector. Since the components of $\vec{W}(M)$ are proportional to the cofactors of dx, dy, dz in the determinant (11.2.1), we have in case (b):

$$(12.7.1)\qquad Ax + By + Cz = 0.$$

This shows that $\vec{r}$ and $\vec{W}(M)$ are orthogonal whenever $\vec{W}(M) \neq 0$. Thus assumption (b) is untenable and (12.7) follows.

(12.8). <u>Remark</u>. The critical points not on the equator are determined by two relations $A(x,y,z) = B(x,y,z) = 0$. For owing to $z \neq 0$, the relation (12.7.1) implies that then also $C = 0$.

(12.9) Suppose that M(x,y,z) is a critical point not on the equator and let $M^*(\xi,\eta,1)$ be the projection πM of M on the plane Π. Since the coordinates of the two points are proportional A, B will vanish at one of the two points whenever it vanishes at the other. Now at M^*, A and B reduce to $P(\xi,\eta)$ and $Q(\xi,\eta)$. Thus M is a critical point on the sphere not on the equator when and only when its projection πM is one on Π.

(12.10) Example. Consider the differential system

(12.10.1)
$$\frac{dx}{dt} = x^2 + y^2 - 1$$
$$\frac{dy}{dt} = 5(xy - 1)$$

discussed by Poincaré (Oeuvres, I, p. 66). In homogeneous form it becomes

(12.10.2)
$$\begin{vmatrix} dx, & dy, & dz \\ x, & y, & z \\ x^2 + y^2 - z^2, & 5(xy-z^2), & 0 \end{vmatrix} = 0$$

or after expansion

(12.10.3)
$$-5z(xy-z^2)dx + z(x^2+y^2-z^2)dy + (4x^2y-y^3-5xz^2+yz^2)dz = 0.$$

The critical points on the equator $z = 0$ correspond to the intersections of the lines $y = 0$, $y = \pm 2x$ with the sphere. They are the six points $(\pm 1, 0, 0)$, $(\frac{\pm 2}{\sqrt{5}}, \frac{\pm 2}{\sqrt{5}}, 0)$ and they are antipodal in pairs.

As already noted the critical points not on the equator are determined by

$$z(xy-z^2) = 0, \qquad z(x^2+y^2-z^2) = 0,$$

or since $z \neq 0$ by

$$xy = z^2 = x^2 + y^2.$$

Since $x^2 + y^2 - xy$ is a definite quadratic form it vanishes only for $x = y = 0$ which is manifestly ruled out. Thus all the critical points are on the equator.

Referring to (12.8), the system (12.10.1) has no finite critical points whatever, and hence it can have no closed characteristics (10.9).

13. Our next objective is to introduce an index for the vector distribution on the sphere. This will be done by means of certain local parametrizations which we must first discuss.

(13.1) Setting $F = x^2 + y^2 + z^2 - 1$, at any point $M(x_0, y_0, z_0)$ on the sphere, one of the partials $\frac{\partial F}{\partial x_0}$, ..., will not vanish. Let it be, for instance, the last. Then in the neighborhood of M the sphere has the representation

(13.1.1) $$z - z_0 = \varphi_1(x-x_0, y-y_0).$$

In more general form we may say that in the neighborhood of M there is a parametric representation which we will call <u>regular at</u> M:

(13.1.2) $$x - x_0 = f_1(u,v), \qquad y - y_0 = g_1(u,v),$$
$$z - z_0 = h_1(u,v),$$

where the Jacobian matrix

$$\left\| \begin{matrix} \frac{\partial f_1}{\partial u}, & \frac{\partial g_1}{\partial u}, & \frac{\partial h_1}{\partial u} \\ \frac{\partial f_1}{\partial v}, & \frac{\partial g_1}{\partial v}, & \frac{\partial h_1}{\partial v} \end{matrix} \right\|_{(0,0)}$$

is of rank two. For if we set $x - x_0 = u$, $y - y_0 = v$ then the Jacobian of the first two columns of the system resulting from (13.1.1) is unity.

Consider a general parametric representation (13.1.2). If say the determinant of the first two columns does not vanish, one may solve for u, v as

(13.1.4) $$u = \psi_1(x-x_0, y-y_0), \quad v = \chi_1(x-x_0, y-y_0)$$

and hence (13.1.2) is a topological mapping of a neighborhood of M on the sphere on a small circular region of center at the origin in the (u,v) plane.

(13.2) Needless to say the regular representation (13.1.2) is far from unique. In fact if

(13.2.1) $$u' = \psi_1(u,v) \qquad v' = \chi_1(u,v)$$

is any transformation regular at the origin then upon substituting in (13.1.2) there is obtained a new representation regular at M:

(13.2.2) $$x - x_0 = f_1^*(u',v'), \ \ldots .$$

Let us prove the converse: all representations regular at M are deducible from one another in the manner just described. Suppose in fact merely that (13.1.2) and (13.2.2) are two such representations and let say

$$\left(\frac{\partial(f_1,g_1)}{\partial(u,v)}\right)_{(0,0)} \neq 0.$$

It is well known that this is a n. a. s. c. in order that dx, dy be linearly independent at M. Hence

$$\left(\frac{\partial(f_1^*,g_1^*)}{\partial(u',v')}\right)_{(0,0)} \neq 0.$$

From this follows that the first two relations of (13.1.2) and of (13.2.2) define regular transformations. Hence the system

$$f_1(u,v) = f_1^*(u',v'), \qquad g_1(u,v) = g_1^*(u',v')$$

can be solved in the form (13.2.1), yielding a regular transformation $(u,v) \longrightarrow (u',v')$.

(13.3) Remark. We have chosen representations regular at M, with M corresponding to $u = v = 0$. This is clearly unnecessary: one may merely require to have a representation $x = f_1(u)$, ..., with $M^*(u_0,v_0)$ corresponding to M, where f_1, ..., are holomorphic at M* and have a Jacobian matrix of rank two at that point.

(13.4) A noteworthy regular representation is the following. Suppose M(x,y,z) not on the equator and let OM intersect Π : z = 1, at M*(ξ,η,1), so that M* = πM. Then

$$\frac{\xi}{x} = \frac{\eta}{y} = \frac{1}{z} = \varepsilon\sqrt{\xi^2+\eta^2+1} = \varepsilon\rho$$

where ε = ±1 and ε has the sign of z. Hence

$$(13.4.1) \qquad x = \frac{\varepsilon\xi}{\rho}, \quad y = \frac{\varepsilon\eta}{\rho}, \quad z = \frac{\varepsilon}{\rho} .$$

This representation is holomorphic and with a Jacobian matrix of rank two whatever M*(ξ,η). The points M = π^{-1}M* are merely any points not on the equator. Hence (13.4.1) is a representation regular outside of the equator.

(13.5) Another interesting regular transformation is obtained by the stereographic projection ω of the sphere S from its north pole N(0,0,1) on the equator. It will be recalled incidentally that ω maps S - N conformally on the equatorial plane. Analytically if M is the point (x,y,z) ≠ N and NM intersects z = 0 at (ξ,η,0) then

$$(13.5.1) \quad \frac{x}{\xi} = \frac{y}{\eta} = 1 - z; \quad \xi = \frac{x}{1-z}, \quad \eta = \frac{y}{1-z} .$$

By means of the equation of the sphere one obtains

$$(13.5.2) \quad x = \frac{2\xi}{\xi^2+\eta^2+1}, \quad y = \frac{2\eta}{\xi^2+\eta^2+1}, \quad z = \frac{\xi^2+\eta^2-1}{\xi^2+\eta^2+1}$$

This representation is readily shown to be regular throughout S - N.

(13.6) Let us apply an affine transformation preserving the origin:

$$x' = a_{11}x + a_{12}y + a_{13}z, \qquad y' = a_{21}x + \dots,$$
$$z' = a_{31}x + \dots,$$

where the determinant $|a|$ of the matrix $a = \| a_{ij} \|$ does not vanish. Corresponding to a parametrization (13.1.2) there will be a similar one expressing x', ..., as holomorphic functions of u,v. Now if J, J' are the corresponding Jacobian matrices such as (13.1.3) in the two representations at any point M then $J' = Ja$. Hence if one of them is of rank two at M so is the other and therefore if one of the two representations is regular at M so is the other. In other words to obtain a regular representation we may operate with the coordinates x', ..., in the same way as with the initial coordinates.

Applications. (13.6.1) In the regular transformation of (13.4) one may assume that the equatorial plane is any diametral plane Π_1, while Π is a plane tangent to the sphere parallel to Π_1. For one may always choose rectangular axes such that Π, Π_1 are the planes $z = 1$, $z = 0$.

(13.6.2) In the rectangular transformation of (13.5) one may assume that N is any point of the sphere whatsoever for one may always choose axes such that N is the point $(0,0,1)$.

(13.7) Remark. It may be observed that the stereographic projection of the southern hemisphere H' from the north pole N yields a regular representation of the former on a circular region. Indirectly also it yields a regular representation of the initial plane x,y of the system (1.1) on a circular region. This representation will often be found useful in studying the complete system of characteristics.

14. (14.1) Take now an arbitrary transformation (13.1.2) regular at $M(x_0,y_0,z_0)$ and suppose that M corresponds to $M^*(u_0,v_0)$. Denote also by ω the transformation $M \to M^*$. For later purposes it is convenient to

denote by $V_x, \ldots$, the components $\frac{dx}{dt}, \ldots$, of $\vec{V}(M)$. Consider also $\frac{du}{dt}, \frac{dv}{dt}$ as the components of a vector $\vec{V}^*(M^*)$ in the (u,v) plane Π and accordingly write them V_u^*, V_v^*. Then (12.6.1) yields

$$(14.1.1) \quad \begin{cases} V_x = \dfrac{\partial f_1}{\partial u} V_u^* + \dfrac{\partial f_1}{\partial v} V_v^* = Bz - Cy, \\ V_y = \dfrac{\partial g_1}{\partial u} V_u^* + \dfrac{\partial g_1}{\partial v} V_v^* = Cx - Az, \\ V_z = \dfrac{\partial h_1}{\partial u} V_u^* + \dfrac{\partial g_1}{\partial v} V_v^* = Ay - Bx. \end{cases}$$

The relation (12.2.2) assumes now the form

$$(14.1.2) \qquad xV_x + yV_y + zV_z = 0$$

which merely states again that $\vec{V}(M)$ is in the plane tangent to S at M. If we consider (14.1.1) as a linear system to express V_u^*, V_v^*, in terms of $V_x, \ldots$, then since x,y,z are not all zero, (14.1.2) shows that the system is compatible. Since the coefficient matrix at M, which is merely the Jacobian matrix of f_1, g_1, h_1 taken at M, is of rank two we have in (14.1.1) a non-singular affine transformation from the space of the vectors $\vec{V}^*$ in the plane Π to the space of the vectors $\vec{V}(M)$ in the tangent plane at M. Hence

(14.1.3) $\vec{V}^*(M^*)$ <u>is zero when and only when</u> $\vec{V}(M)$ <u>is zero</u>.

In other words if M is a critical point for the vector distribution $\vec{V}(M)$ on S then M^* is one for the vector distribution $\vec{V}^*$ in the plane Π and conversely.

The explicit solution of (14.1.1) for V_u, V_v yields two relations valid in the neighborhood of M^* and of form

$$(14.1.4) \qquad \frac{du}{dt} = R(u,v), \qquad \frac{dv}{dt} = S(u,v)$$

where R, S are holomorphic at M^* and $R = S = 0$ there when and only when $\vec{V}(M) = 0$.

Now within a suitable small circle of center M^* in the plane Π one may apply to (14.1.4) the same arguments as in the earlier sections for systems valid in the whole plane. In particular let M be critical. Then M^* is likewise critical and hence an isolated critical point like M itself. If M^* is an elementary critical point: focus, ..., then M is said to be elementary and called a focus, ... on S. The Index (M^*) for the vector distribution $\vec{V}^*$ at M^* is by definition the Index (M) of M for the vector distribution $\vec{V}(M)$ at M. Since the effect of changing the regular representation at M is merely to apply to the plane Π a transformation of coordinates regular at M, the properties just defined are independent of the regular representation in terms of which they have been described.

(14.2) Consider in particular the special regular representation of (13.4). In the notations there chosen the analogue of (14.1.4) becomes

$$(14.2.1) \qquad Q(\xi,\eta)d\xi - P(\xi,\eta)d\eta = 0.$$

Hence the critical points not on the equator are those projected by π into those of (14.2.1), i. e. into those of (1.1), a property already proved in (12.8). Moreover the Index (M) of a critical point not on the equator is the same as Index (M^*), the index of its projection M^* on the plane Π. In other words the projection from sphere to plane preserves the nature of the critical points not on the equator as well as their index. Notice also that as a consequence of the preceding remarks, if M is a non-equatorial critical point then so is its antipode M' and both have the same index.

Let us show that this last property holds also when M is a critical point on the equator. In fact the argument just made is valid then provided that we project on

$x = 1$ when $x \neq 0$ at M, or on $y = 1$ when $x = z = 0$. The only difference is that the analogue of (14.1.2) then obtained will have no simple relation to (1.1). The conclusion will however be the same: Index (M) = Index (M').

We may therefore state:

(14.3) The critical points on the sphere are distributed in antipodal pairs and the indices in each pair are equal.

15. The sum of the indices of all the critical points on the sphere S is called the index of S, written Index (S). We shall prove the following property, due to Poincaré:

(15.1) The index of a sphere is always equal to two.

(15.2) It may be observed that Poincaré proved a more general result. Namely one may introduce vector distributions on any closed surface Φ, (compact two dimensional analytical manifold). When the critical points are isolated one may define the index (Φ) and Poincaré has shown that Index (Φ) = $\chi(\Phi)$, the Euler-Poincaré characteristic (number of triangles - number of sides + number of vertices in any triangulation of Φ). The characteristic of an orientable surface of genus p, Φ_p, being $2 - 2p$, we have then Index (Φ_p) = $2 - 2p$. The sphere is merely Φ_0 and so Index (S) = 2 which is (15.1).

(15.3) The preceding results are closely related to another question. A vector distribution say on Φ defines a continuous deformation of Φ into itself, whose fixed points are the critical points. The author's fixed point formula (algebraic count of the fixed points) yields then the index properties as a very special case.

(15.4) Proof of (15.1). Choose a coordinate sphere whose north pole N (0,0,1) is not a critical point of the vector distribution or field $\mathcal{F}$ on S. Let Π denote the equatorial plane $z = 0$ and ω the stereographic projection of S - N onto Π from N. Then ω yields a regular representation at every point of S - N in terms of the coordinates on Π. We use the notations of (14) and denote the projections of M, ..., in S - N by M*, ... In particular, there results a field $\mathcal{F}$* in Π and, if $A_1, \ldots, A_s$ are the critical points of $\mathcal{F}$ (none is N), then $A_1^*, \ldots, A_s^*$ are those of $\mathcal{F}$*. Since ω preserves the index

$$(15.5)\quad \text{Index }(S) = \sum \text{Index }(A_h) = \sum \text{Index }(A_h^*).$$

Since the critical points of $\mathcal{F}$ are isolated and do not include N, we may choose an $\varepsilon > 0$ so small that the spherical cap $\Omega : z \leq 1 - \varepsilon$ contains all the A_h. Let Γ be the boundary circumference of Ω. As a consequence the circular region Ω* will contain all the A_h^*. Hence

$$(15.6)\quad \text{Index }(\Gamma^*) = \sum \text{Index }(A_h^*) = \sum \text{Index }(A_h) = \text{Index}(S).$$

Therefore the proof of the theorem reduces to showing that

(15.7) <u>For a sufficiently small</u> ε : Index $(\Gamma^*) = 2$.

This will be done in two steps. We shall show that Index (Γ) hence Index (Γ*), is independent of the field $\mathcal{F}$, then calculate the index for a suitably chosen simple field.

(15.8) <u>Proof of the independence of the index from the field</u>. Supposing now M on the circle Γ we shall modify the vector $\vec{V}(M)$ without changing the index so that this vector takes a value independent of the field. This will prove the property in question. Referring to (14.1) where $V_x, \ldots$, are now the components of $\vec{V}(M)$, we may define $\vec{V}^* = \omega V(M)$, i. e., solve (14.1.1) for the components V_u^*, V_v^* of $\vec{V}^*$, when and only when $V_x, \ldots$, satisfy the compatibility relation (14.1.2), or when and only when $\vec{V}(M)$ is in the tangent plane at M. To be certain that $\vec{V}(M)$ remains in the tangent plane at M throughout its variation, it is most convenient to have recourse to the relation

$$(15.8.1) \qquad \vec{V}(M) = \vec{r} \times \vec{W}(M) = \overrightarrow{OM} \times \vec{W}(M)$$

where initially $\vec{W}(M)$ is as in (12.5). For whatever $\vec{W}(M)$ the vector $\vec{V}(M)$ determined by means of (15.8.1) is orthogonal to $\overrightarrow{OM}$ and hence in the tangent plane at M. Thus all that is necessary now is to vary $\vec{W}(M)$ in a suitable way and to define $\vec{V}(M)$ at M by means of (15.8.1). This is the procedure followed below.

Let then Λ be the northern spherical cap $1 - \varepsilon \leq z \leq 1$ and corresponding to any two points M, M' on Λ and to $0 \leq k \leq 1$ form the vector

$$\vec{V}_k(M,M') = \overrightarrow{OM} \times \{k\vec{W}(M') + (1-k)\vec{W}(M)\}$$

which is manifestly continuous whatever k and whatever M, M' on Λ. Now whatever k and for M = M' = N the vector reduces to $\vec{V}(N) \neq 0$ since N is not critical. Hence for ε sufficiently small and k arbitrary on the unit-segment, $\vec{V}_k(M,M') \neq 0$. Choose in particular M' at N and M on Γ_1. Thus $\vec{V}_k(M,N)$ is a vector in the tangent plane at M and $\neq 0$ whatever k. It follows that $\vec{V}_k^*(M,N) = \omega\vec{V}_k(M,N)$, which we consider as applied to M^* in Π, is $\neq 0$ whatever k. We may therefore take the index $I_k(\vec{V}^*)$ of that vector along Γ^*; it is a continuous function of k and since it is an integer it is constant. Now for k = 0, $\vec{V}_k(M,N)$ reduces to $\vec{V}(M)$ and for k = 1 to $\vec{r} \times \vec{W}(N)$. Hence in taking Index (Γ^*) we may, for ε small enough, replace along Γ the vector $\vec{W}(M)$ by $\vec{W}(N)$, its value at the north pole N.

Let now α be the angle which $\vec{W}(N)$ makes with the direction ON. Since $\vec{V}(N) \neq 0$ we may assume that $0 < \alpha < \pi$. Modify now $\vec{W}(N)$ without changing its length so that $\alpha \to \frac{\pi}{2}$ monotonely. As a consequence $\vec{r} \times \vec{W}(N)$, for ε small enough and for M on Γ is never zero. Hence the same argument as above shows that under these conditions Index (Γ^*) remains the same. Since the length of $\vec{W}(N)$ and its actual position in a meridian plane through ON do not affect the index, we may finally suppose that $\vec{W}(N)$ is the fixed vector (1,0,0). Since the resulting index is the same whatever the field, Index (Γ) and hence Index (S) is independent of the field.

(15.9) Construction of a field whose index sum is two. Consider the special system (1.1):

$$\frac{dx}{dt} = y, \qquad \frac{dy}{dt} = -x. \tag{15.10}$$

The corresponding homogeneous form is

$$-xzdx + yzdy - (x^2+y^2)dz = 0.$$

Hence the only critical points are $(0,0,\pm 1)$, and both correspond to the critical point of (15.10) at the origin. This critical point is elementary and is a center since the characteristics are $x^2 + y^2 = k$. Hence its index is one and the index sum for the whole sphere is two.

This completes the proof of Poincaré's index theorem for the sphere.

16. Effective construction of a system of characteristics illustrated on an example. Let us return to Poincaré's example considered in (12.8) and endeavor to describe the complete system of the characteristics on the sphere. Owing to the central symmetry of the system it is only necessary to consider one of the two hemispheres. We shall assume that it is the southern hemisphere and consider its image as a circular region.

(16.1) It is first necessary to determine the nature of the critical points. There had been found six, in three antipodal pairs, all on the equator. Let us select the following three, one in each antipodal pair:

$$A(1,0,0), \qquad B\left(\frac{1}{\sqrt{5}}, \frac{2}{\sqrt{5}}, 0\right), \qquad C\left(\frac{1}{\sqrt{5}}, \frac{-2}{\sqrt{5}}, 0\right).$$

Critical point A. Choose as plane Π the plane $x = 1$ and project from the center onto that plane. As a consequence (12.8.3) becomes

$$z(1+y^2-z^2)dy + (4y-y^3-5z^2+yz^2)dz = 0.$$

The critical point A is imaged into the origin of the y,z plane and it is the critical point of the system

$$\frac{dy}{dt} = -4y + 5z^2 - yz^2 + y^3,$$

$$\frac{dz}{dt} = z + z(y^2-z^2).$$

The first approximation is

$$\frac{dy}{dt} = -4y, \qquad \frac{dz}{dt} = z.$$

The critical roots at the origin are -4, 1. Hence the origin is a saddle point and so is A on the sphere. Hence Index (A) = -1. Notice that the tangents at A to the two characteristics tending to the saddle point A are y = 0, z = 0, i.e., the line at infinity and the x axis.

Critical point B. Apply the affine transformation of variables $x \longrightarrow x$, $y \longrightarrow y + 2x$, $z \longrightarrow z$. As a result the point B will acquire the coordinates $(\frac{1}{\sqrt{5}},0,0)$. Applying the above change of coordinates to (12.8.3), making x = 1, dx = 0, then preserving only the first degree terms in y, z there is obtained

$$5zdy - 4ydz = 0.$$

Hence the first approximation is now

$$\frac{dy}{dt} = 4y, \qquad \frac{dz}{dt} = 5z.$$

The characteristic roots are 4,5. Hence B is a node and Index (B) = +1.

Critical point C. This time the change of variables is $x \to x$, $y \to y - 2x$, $z \to z$, but otherwise the calculation is the same leading to the same result; C is also a node and Index (C) = +1.

There are then on the equator 4 nodes and 2 saddle points and the sum of the indices is 2, as it should be.

(16.2) Let us mark (fig. 14) the various critical points on the boundary of the circular region on which the system of characteristics is to be represented. The basic differential equations

$$(16.2.1) \quad \frac{dx}{dt} = x^2 + y^2 - 1, \qquad \frac{dy}{dt} = 5xy - 5,$$

which we repeat for convenience, will enable us to determine the directions of approach. Since $\frac{dx}{dt} > 0$ for x large on the x axis, the arrows along A'A (near A,A') must be pointed out as shown. Since AB, for instance, is a characteristic and A is a saddle point, with x axis and AB as the tangents to the two characteristic arcs tending to A, the arrow along AB must point from A to B. Since

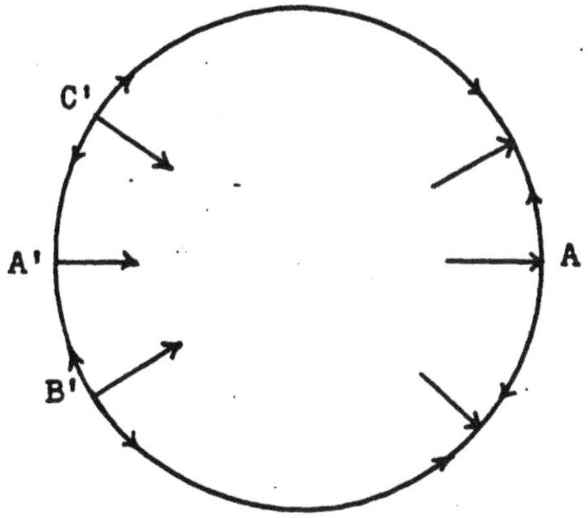

Figure 14

B is a node all the characteristics from B must point the same way, etc. Proceeding thus there is obtained the full scheme of pointers indicated in fig. 14.

Let Γ denote the circle $x^2 + y^2 = 1$ and Δ the hyperbola $xy = 1$. From (16.2.1) we learn that along a characteristic x decreases inside Γ and increases outside Γ, while y decreases outside Δ and increases in its two interior regions. This situation is described by horizontal and vertical pointers in fig. 15. Observe also that Γ is

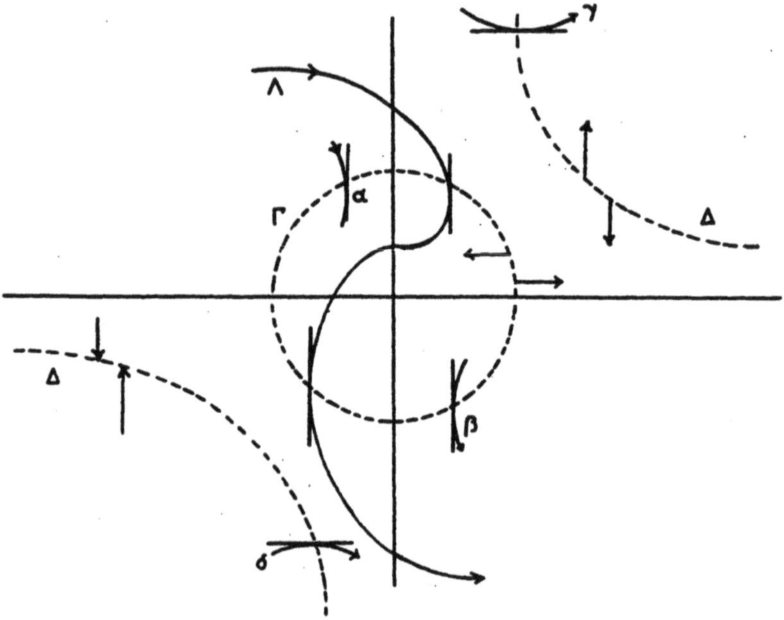

Figure 15

the locus of the points where the tangents to the characteristics are vertical and Δ the locus of the points where they are horizontal. The only crossings of Γ, Δ by characteristics consistent with these various properties are as indicated by the arcs α, β, γ, δ, in fig. 15.

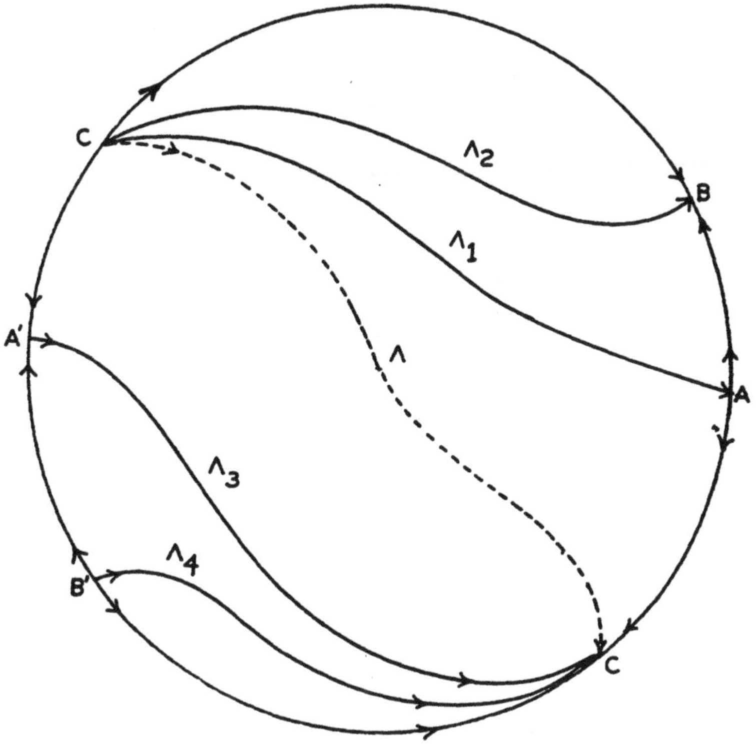

Figure 16

If we recollect that within Γ the slopes $\frac{dy}{dx}$ of the characteristics are always ≥ 0, we find that the characteristics crossing Γ have the general form of Λ in fig. 15. Returning now to the circular region of fig. 14 we find that Λ assumes the general form indicated in fig. 16. From this follows that the two characteristics issued from the two saddle points other than the equator are of the types Λ_1, Λ_2 of fig. 16. The other basic types are then represented by Λ_3, Λ_4.

§6. THE LIMITING SETS AND LIMITING BEHAVIOR OF CHARACTERISTICS

17. To complete the examination of the behavior of the characteristics in the large we still require information regarding their performance as $t \to \pm\infty$. This information will be obtained by means of the so-called limiting sets.

The basic differential system is (1.1) extended to the sphere S, i.e. (12.6.1), and a characteristic γ is a solution of (12.6.1). We recall, however, that on S the equator $z = 0$ (the line at infinity) is made up of characteristics and critical points. Hence no characteristic crosses from one hemisphere to the other. More precisely, every characteristic remains in a spherical cap, i. e. in a closed circular region. This will enable us, in local arguments, to pass to approximating cartesian coordinates, i. e., in the last analysis to assume that the basic system is still (1.1). The "local" terminology will, then, be that of the plane.

In point of fact the whole argument to follow is valid for any bounded characteristic in the plane without any necessity to pass through the medium of the sphere.

Before dealing with the main argument it will be necessary to discuss a certain number of preliminary properties.

(17.1) Lemma. *Let the Jordan curve J on the sphere S possess an analytical arc λ and let μ be an analytical arc on S meeting J at a single point P of λ where λ and μ are not tangent. Then two points Q, R of μ which are separated by P are also separated by J. That is to say Q is in one of the two regions in which J decomposes S, and R in the other.* (Figure 17.)

Take a point N in S-J and project S stereographically from N on a plane Π. This will reduce the lemma to a similar property for the plane. To avoid multiple changes of notation suppose now that J, ..., are all in Π. Since λ is analytical there is a conformal mapping φ of J and its interior on the closed circular region, $x^2 + y^2 \leq 1$ where φ may be extended analytically and topologically across λ. Since Q and R may be replaced by points arbitrarily near P, we may assume μ in the region in which φ operates. This reduces at once the plane problem to the same in which J is now the circle $x^2 + y^2 = 1$ and μ crosses J without contact. Let r be the radius vector, and let $x = f(t)$, $y = g(t)$ be an analytical parametrization of μ. Now

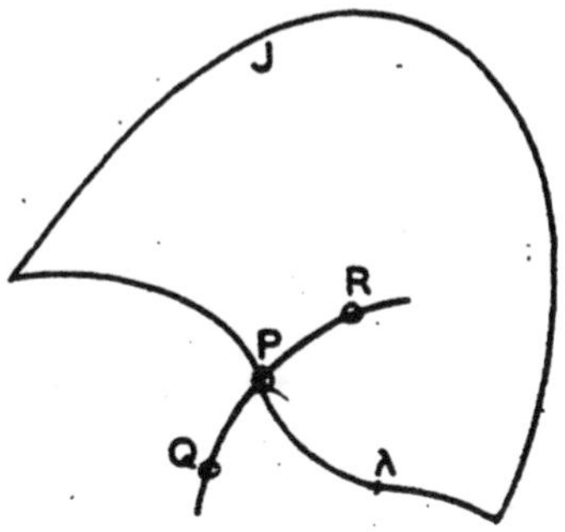

Figure 17

$$\frac{rdr}{dt} = \frac{xdx}{dt} + \frac{ydy}{dt} \neq 0$$

at P since otherwise λ and μ would be tangent there. Hence we may assume μ so small that $\frac{dr}{dt}$ has a fixed sign as t increases and the point M $(f(t), g(t))$ follows μ. If P corresponds to the value $t = t_0$, then as t passes t_0 the radius vector will pass monotonely through the value one and M will pass from the interior of the circle to its exterior or vice versa. Hence P separates Q from R.

The remaining preliminary properties to be considered refer directly to characteristics.

(17.2) <u>The characteristics which traverse a characteristic rectangle all describe it in the same direction.</u>

This is implicit in the definition of the rectangles.

(17.3) <u>Arcs without contact</u>. This convenient concept was introduced by Poincaré and repeatedly utilized by him and by Bendixson in connection with the questions here discussed. It will be sufficient for our purpose to consider a closed analytical arc λ. If λ satisfies $f(x,y) = 0$, we will say that λ is without contact whenever it contains no critical points and at any interior point P of λ the characteristic γ through P has the property that as a point M of γ follows γ and passes through P the value of f at M changes its sign. Geometrically this is well known to mean that γ has a contact of even order (ordinary crossing, inflection point, etc.) with λ at P. However, for our purpose it is more convenient to have the analytical formulation.

(17.4) Example: if γ is a characteristic (not a critical point), then a sufficiently short segment σ, transverse to γ at any point M which is the midpoint of σ, is an arc without contact.

(17.5) A noteworthy <u>characteristic rectangle</u> (generalization of the earlier configuration of the same name) is associated with the arc without contact λ. Since λ is analytical it is rectifiable. Let 2α be its length, A its midpoint, s the arc length counted positively along λ in a certain direction. If P is any point of λ and γ the characteristic through P, let γ be made to correspond to the solution $(x(t), y(t))$ of (1.1) such that $(x(0), y(0))$ is the point P. Changing if necessary f into $-f$, we can find a $\tau > 0$ such that on any γ the function $f \geq 0$ for $0 \leq t \leq \tau$, and $f \leq 0$ for $-\tau \leq t \leq 0$. Consider now s, t as rectangular coordinates of an Euclidean plane, Π, and let R^* be the rectangle in Π determined by $|s| \leq \alpha$, $|t| \leq \tau$. Define a mapping φ of R^* whereby the point M^* (s,t) goes into the point M $(x(t), y(t))$ of the characteristic γ issued from the point P of λ at a distance s from A. The mapping φ is manifestly continuous and one-one. Therefore since R is compact, φ is topological. We shall refer to the image $R = \varphi R^*$ as a <u>characteristic rectangle of median line</u> λ.

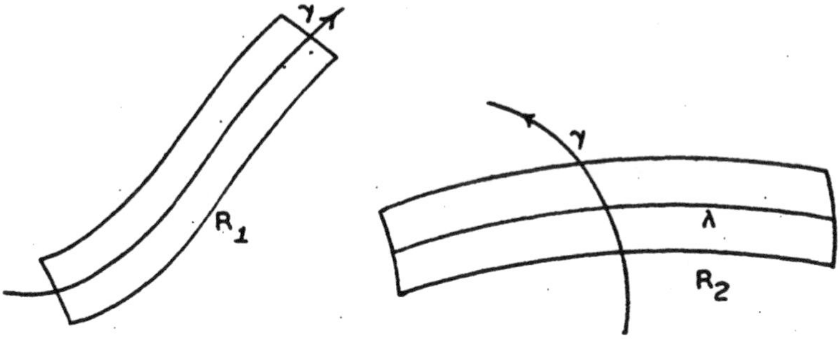

Figure 18

This mild generalization of the earlier characteristic rectangles has all their properties. The distinction occurs primarily in the applications. The earlier characteristic rectangle (R_1 in fig. 18) always has a preassigned axis but may be made very thin around that axis. The present type (R_2 in fig. 18) has a preassigned basic line but may be very thin around that line, i. e. in the transverse direction relative to the characteristics.

(17.6) <u>Lemma</u>. <u>If a characteristic</u> γ, <u>followed as</u> t <u>increases, has with an arc without contact</u> λ <u>three consecutive interior crossings</u> M, M_1, M_2, <u>then</u> M_1 <u>is between</u> M <u>and</u> M_2 <u>on</u> λ.

If the lemma is incorrect, then M_2 is between M and M_1 or M between M_2 and M_1. The first case corresponds to fig. 19, and the second to fig. 20. Choose at all events λ so small that there is a characteristic rectangle related to it as in the two figures. Taking the case of fig. 19, let J be the Jordan curve MSM_1M_2M. Since J does not separate P from Q nor from R, it must not separate Q from R, a violation of the preceding lemma. In the case of fig. 20 let J be the Jordan curve PSQTP.

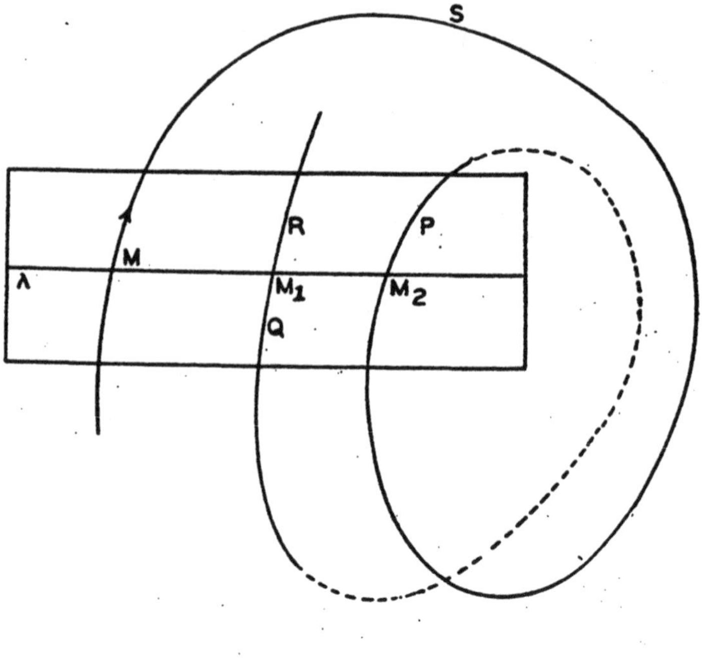

Figure 19

Then J separates M_1 from M_2 since M separates them on λ and λ is not tangent to the analytic (characteristic) arc PSQ at M. On the other hand M_1 and M_2 are manifestly not separated by J. Thus in both cases a contradiction arises and so the lemma is proved.

(17.7) Lemma. <u>Suppose that a critical point</u> O <u>is chosen as the origin and let</u> M (x(t), y(t)) <u>be such that</u>

its coordinates are a solution of (1.1) *corresponding to a characteristic* γ. *Let also* $(r(t), \theta(t))$ *be the polar coordinates of* M. *If as* $t \to \pm \infty$, $M \to 0$, *i. e.* $r \to 0$, *then* $\theta(t)$ *has a limit which may be finite or infinite, i. e.* $\gamma \to 0$ *in a fixed direction or else by spiralling around* 0. (See the complement given in 20.5).

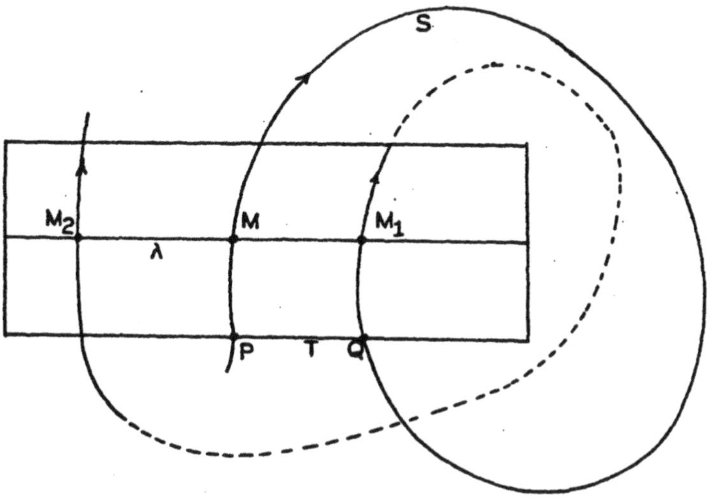

Figure 20

We deduce at once from (1.1)

$$r^2 \frac{d\theta}{dt} = xQ - yP = r^{m+1}(\alpha_{m+1}(\theta) + r\alpha_{m+2}(\theta) + \dots)$$

where $\alpha_h(\theta)$ is a form of degree h in sin θ, cos θ. Let us suppose that $r \to 0$ as $t \to +\infty$. The case $t \to -\infty$ may be taken care of by replacing t by -t. To prove that

$\theta(t)$ has a finite or infinite limit, it is sufficient to show that when it remains bounded it cannot have two distinct limiting values θ_0, θ_1. Suppose that this is the case. One may select a θ_2 between θ_0 and θ_1 such that $\alpha_m(\theta_2) \neq 0$. Thus for r sufficiently small $\frac{d\theta}{dt}$ will have the sign of $\alpha_m(\theta_2)$ and hence the radius vector $L:\theta = \theta_2$ will be crossed by γ always in the same direction. This is, however, impossible since L is crossed by arcs of γ arbitrarily near 0 joining points arbitrarily near θ_0 to points arbitrarily near θ_1 in both possible directions. Thus $\theta(t)$ has a limit.

In an important case it is possible to tell almost by inspection what the directions of approach are. Suppose in fact that

$$(17.8) \quad P = P_m + P_{m+1} + \cdots, \qquad Q = Q_m + Q_{m+1} + \cdots, \qquad m \geq 1,$$

where P_h, Q_h are forms of order h in x,y and one of P_m, Q_m is not identically zero. Then

(17.9) <u>If the form $R(x,y) = yQ_m - xP_m$ is not identically zero, then the directions of approach are those represented by $R = 0$. Thus there are at most</u> $2m+2$ <u>directions of approach and they are opposite in pairs.</u>

We have in fact $\frac{d\theta}{dt}$ as before but with

$$\alpha_{m+1}(\theta) = \frac{1}{r^{m+1}} R(x,y) \neq 0.$$

In addition

$$r\frac{dr}{dt} = r^{m+1} \{\beta_{m+1}(\theta) + r\beta_{m+2}(\theta) + \cdots\}$$

where β_h is like α_h. Along any characteristic γ tending to 0 replace the variable t by $\tau = \int_{t_0}^{t} r^{m-1}dt$. As a consequence, instead of $t \to +\infty$ we may have to consider τ tending to a finite value τ_1. (We shall see, however, that $\tau \to +\infty$ as before.) The new system for γ is now

$$(17.10)\qquad \begin{aligned} \frac{d\theta}{d\tau} &= \alpha_{m+1}(\theta) + r\alpha_{m+2}(\theta) + \dots \\ \frac{dr}{d\tau} &= r\{\beta_{m+1}(\theta) + r\beta_{m+2}(\theta) + \dots\} . \end{aligned}$$

Let r,θ be considered in (17.10) as cartesian coordinates. To the critical points of (1.1) other than the origin there correspond now critical points at distances from the θ axis $\geq \rho > 0$. The new critical points on $r = 0$ arise at the places where $\alpha_{m+1}(\theta) = 0$ and they are discrete. In fact, they are the values of θ corresponding to the directions represented by $R = 0$. Now if $f(t)$ is any solution of

$$\frac{d\theta}{d\tau} = \alpha_{m+1}(\theta),$$

the system $(\theta = f(\tau),\ r = 0)$ represents a solution of (17.10). Hence the arcs of the axis $r = 0$ between critical points are characteristics. It follows that no characteristic γ can tend to a point of such a characteristic from outside $r = 0$. Hence γ can only tend to a critical point of $r = 0$, and it does so as $\tau \longrightarrow +\infty$. This obviously implies (17.9).

The property just proved can readily be verified on the elementary critical points.

18. We come now to the limiting sets. It is upon them that the whole of the subsequent argument will rest.

(18.1) Let γ be a characteristic, $\gamma^+(M)$ the subset of γ consisting of M and of all points of γ traversed after M, $\gamma^-(M)$ the analogue referring to the points traversed before M. We refer to $\gamma^+(M)$ and $\gamma^-(M)$ as the _positive_ and _negative half-characteristics determined by_ M. The closures $\overline{\gamma}^+(M)$, $\overline{\gamma}^-(M)$ give rise through their intersections to two new sets

$$\Lambda^+(\gamma) = \cap\, \overline{\gamma}^+(M), \qquad \Lambda^-(\gamma) = \cap\, \overline{\gamma}^-(M)$$

called the _positive_ and _negative limiting sets_ of γ.

Since $\gamma^+(M)$, $\gamma^-(M)$ are connected, so are their closures: The latter being also compact and non-empty, we have from well-known point-set properties:

(18.2) The limiting sets $\Lambda^+(\gamma)$, $\Lambda^-(\gamma)$ are compact, connected, and non-empty.

Obviously also:

(18.3) If γ is a critical point, then it coincides with its limiting sets.

(18.4) A characteristic such as δ contained in any limiting set $\Lambda^+(\gamma)$, $\Lambda^-(\gamma)$ is known as a separatrix. The knowledge of these curves may be very useful in tracing out the full diagram of the characteristics.

(18.5) Suppose then that P is an ordinary point of say $\Lambda^+(\gamma)$. Thus there is a characteristic δ through P. Construct a characteristic rectangle Ω with P on its axis and interior to Ω and let σ be a small segment transverse to δ at P. Since $P \in \gamma_M^+$, there are points of γ_M^+ arbitrarily near P and the arcs of γ_M^+ through them will contain subarcs in Ω. In particular, these arcs will all cross σ in the same direction and will meet it in points $R_1, \ldots, R_s, \ldots$, met in that order as γ_M^+ is described forward from M. By virtue of Lemma (17.6) the R_s are in the same order on σ. Hence they can have only one limit-point on σ and since P is manifestly such a point, $R_s \to P$. Moreover, on σ the points R_s are all on the same side of P. Or explicitly:

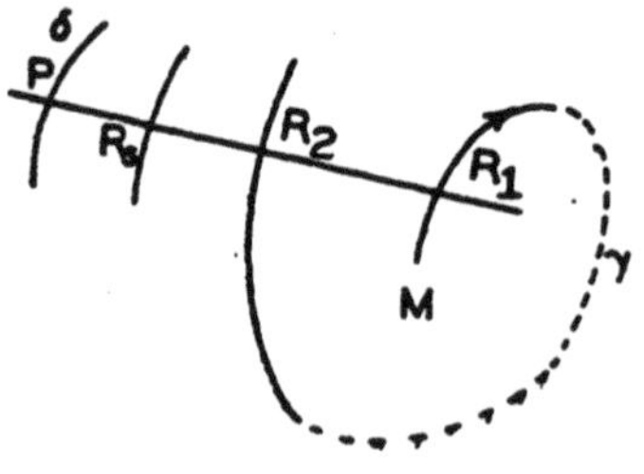

Figure 21

(18.6) If P is an ordinary point of $\Lambda^+(\gamma)$, δ the characteristic through P, σ a suitably small segment transverse to δ at P, then $\gamma^+(M)$ followed forward meets σ, in successive points R_1, R_2, ..., which are ranged in the same order on σ, and tend to P on one side of P on σ.

The notations being as before, let P' be a point of $\Lambda^+(\gamma)$ in Ω and δ' its characteristic. Since δ' meets Ω it crosses σ in a point P" which is a limit-point of the R_s. Hence P" = P, $\delta' = \delta$, and P' is in the axis λ of Ω. Since Ω is a neighborhood of P, we have:

(18.7) Corresponding to any open arc λ of the separatrix δ and point P of λ, there is a neighborhood U of P such that $U \cap \Lambda^+(\gamma) \subset \lambda$.

(18.8) A characteristic which meets a limiting set is contained in that set and hence it is a separatrix.

It is sufficient to consider $\Lambda^+(\gamma)$, and to show that if it meets the characteristic δ then it contains δ. When δ is merely a critical point this is obvious so this case may be dismissed. Let $E = \delta \cap \Lambda^+(\gamma)$. Since δ is a Jordan curve or an arc, if $E \neq \delta$ there is a boundary point P of E in δ. Since E is a closed subset of δ, P is in E and hence in $\Lambda^+(\gamma)$. Identify now P with the point thus designated before. The arc δ^* of δ in Ω (the axis of Ω) consists of limit-points of the arcs of γ^+_M through the points R_q, R_{q+1}, ..., for q above a certain value. Hence $\delta^* \subset \overline{\gamma^+_M}$ whatever M, and therefore $\delta^* \subset \Lambda^+(\gamma)$. Thus P is an interior point and not a boundary point of E. Therefore $E = \delta$, $\delta \subset \Lambda^+(\gamma)$.

(18.9) The possibility that the separatrix $\delta = \gamma$ is not excluded. In that case P must be one of the points R_q. This can only occur when the R_q all coincide with P and γ is closed. Conversely γ closed implies that $\Lambda^+(\gamma) = \Lambda^-(\gamma) = \gamma$ and that all the R_q coincide with P. Notice also that a n.a.s.c. for $\delta = \gamma$ is that the two intersect, i.e. that $\Lambda^+(\gamma)$ meet γ. Since the properties of $\Lambda^-(\gamma)$ follow from those of $\Lambda^+(\gamma)$ by changing t into -t, we may state:

(18.10) A n.a.s.c. for γ to be closed is that one of $\Lambda^+(\gamma)$, $\Lambda^-(\gamma)$ meet γ, and then both coincide with γ.

Thus γ is then a separatrix.

19. We are now in position to give a complete description of the limiting sets and of the related behavior of the characteristics.

(19.1) Theorem. The limiting sets of a characteristic γ fall under the following mutually exclusive categories:

(19.1.1) $\Lambda^{+}(\gamma)$ consists of a single point A which is critical and with increasing time γ spirals toward A or else converges to A in a fixed direction.

(19.1.2) $\Lambda^{+}(\gamma)$ is a closed characteristic δ and with increasing time γ spirals toward δ on one side of δ.

(19.1.3) γ is contained in a 2-cell E^2 whose boundary is $\Lambda^{+}(\gamma)$. The latter is a graph whose sides are characteristics and whose vertices are critical points. Moreover the sides of the graph are separatrices which tend in each sense to one of the vertices in a definite direction. The characteristic γ spirals towards $\Lambda^{+}(\gamma)$ and any characteristic which starts from a point of E^2 sufficiently near to $\Lambda^{+}(\gamma)$ spirals towards the latter in the same direction as γ. This last property holds also for (19.1.1) and all the characteristics starting sufficiently near A, and likewise for (19.1.2) and those starting sufficiently near δ on the same side as γ.

(19.1.4) $\Lambda^{-}(\gamma)$ is of the same nature as $\Lambda^{+}(\gamma)$ save that the spiralling and tending to a limit occur as $t \to -\infty$.

Figure 22 illustrates the case of a graph and the related behavior of γ.

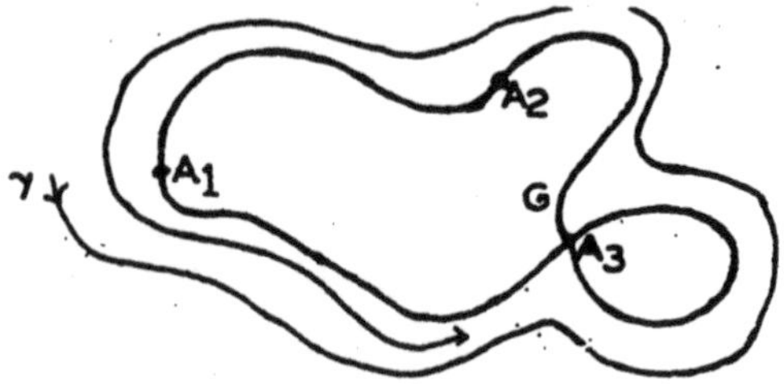

Figure 22

(19.2) Part (19.1,4) is obvious so that it is sufficient to examine the other parts. The simpler cases will be taken up first at some length.

20. (20.1) Suppose first $\Lambda^+(\gamma)$ finite. Since it is connected, it must consist of a single point A. The characteristic through A is in $\Lambda^+(\gamma)$, hence it is A itself. Thus A is a single critical point. Combining this with lemma (17.7) we find that under the circumstances the behavior of γ and $\Lambda^+(\gamma)$ conforms with theorem (19.1).

(20.1.1) It may be observed that obviously when γ tends to a critical point A then $\Lambda^+(\gamma)$ consists of A alone. Similarly when γ tends to A with $t \to -\infty$, then $\Lambda^-(\gamma) = A$.

(20.2) Suppose now that $\Lambda^+(\gamma)$ contains no critical points. Since the set is not empty it contains ordinary points and the characteristics through them are separatrices. Let δ be one of them. Suppose $\Lambda^+(\delta) - \delta$ nonempty and let P be a point of the set. By hypothesis the characteristic δ_1 through P is distinct from δ. Since P is in the closure of subsets of δ it is in $\bar{\delta}$ and hence in the closed set $\Lambda^+(\gamma)$ which contains δ. Since $P \in \Lambda^+(\delta)$ there are points of δ, i. e. of $\Lambda^+(\gamma)$, arbitrarily near P and not in δ_1. As this contradicts (18.7), we must have $\Lambda^+(\delta) = \delta$ and hence δ is closed. Thus every characteristic in $\Lambda^+(\gamma)$ is closed.

Suppose that $\Lambda^+(\gamma)$ contains two distinct closed characteristics δ, δ'. Since δ, δ' are closed and disjoint $\Lambda^+(\gamma)$ cannot be connected, contrary to (18.7).

To sum up:

(20.3) Whenever $\Lambda^+(\gamma)$ contains no critical point, it consists of a single closed characteristic.

A noteworthy incidental consequence is the following proposition due to Poincaré:

(20.4) Theorem. A closed region Ω which is free from critical points and contains a half-characteristic contains also a closed characteristic.

At the cost of changing t into -t we may suppose that the half-characteristic is $\gamma^{+}(M)$. Then $\Lambda^{+}(\gamma) \subset \overline{\gamma}^{+}(M) \subset \Omega$. Hence $\Lambda^{+}(\gamma)$ contains no critical point and so it is a closed characteristic.

As an application which we shall require later we prove the following complement of lemma (17.7). Using the same notations as in the lemma and its proof we may state:

(20.5) <u>If there is a single spiralling characteristic</u> γ <u>tending to the critical point</u> 0 <u>as</u> $t \to +\infty$ [<u>as</u> $t \to -\infty$] <u>then every characteristic passing sufficiently near</u> 0 <u>will likewise</u> $\to 0$ <u>as</u> $t \to +\infty$ [<u>as</u> $t \to -\infty$] <u>and this in the same direction as</u> γ. <u>Thus</u> 0 <u>behaves like a focus</u>.

It is sufficient to take $t \to +\infty$. Choose a circle Γ of center 0 and radius so small that: (a) it contains no other critical point other than 0; (b) it is crossed by γ. Now a closed characteristic σ in Γ must surround 0 and hence must be crossed by γ which is ruled out. Hence there will be no closed characteristics interior to Γ.

Referring now to (17.7) we can find a ray OL containing a segment OA traversed by all the characteristics crossing it in the same direction. Since γ spirals around 0 it will meet OA. Let P_1 be a crossing of OA by γ, P_2, P_3, ..., the consecutive crossings of L by γ beyond P_1. They must follow in the order $P_1P_2P_3 \ldots 0$ and $\to 0$, hence they are all on OA. For suppose P_2 beyond A on OL. Then P_3 is between 0 and P_1 or beyond P_2. In the former case the arcs P_1P_2 and P_2P_3 of γ will necessarily cross, which is ruled out. Hence P_3 is beyond P_2, etc. As a consequence $\{P_n\}$ does not tend to 0 contrary to the assumption that $\gamma \to 0$. Thus the order is $AP_1P_2P_3O$. Consider now the Jordan curve J_n made up of the arc P_nP_{n+1} of γ and of the segment P_nP_{n+1} and let U_n be the region of the sphere bounded by J_n and containing the point 0.

It is at once apparent that $U_{n+1} \subset U_n$. Moreover as $P_n \to 0$ likewise $J_n \to 0$ and hence $U_n \to 0$. Hence we may choose P_n such that $\overline{U}_n$ is interior to the circle Γ. Let us suppose that P_1 is so chosen that $\overline{U}_1$ is already interior to Γ. Then every $\overline{U}_n$ will have the same property. As a consequence $\Omega_n = \overline{U_n - U_{n+1}}$ will be free from critical points and from closed characteristics.

Consider in particular Ω_1, the closed region bounded by the arc $P_1P_2P_3$ of γ and by the segment P_1P_3 (the shaded region in fig. 23). Let now γ' be any characteristic with points in $\overline{U}_1$, and hence in some Ω_n. For simplicity we may suppose that γ' has points in Ω_1. Let M be such a point. Then $\gamma'^{+}(M)$ will have to cross P_2P_3 as $t \to +\infty$. For otherwise $\gamma'^{+}(M) \subset \Omega_1$, and since Ω_1 is free from critical point it contains a closed characteristic δ, contrary to assumption. Hence $\gamma'(M)$ crosses P_2P_3. Similarly $\gamma'^{-}(M)$ crosses P_1P_2. Thus γ' is merely a characteristic issued from a point of P_1P_2 and it must reach P_2P_3 in Ω_1. Similarly beyond P_2P_3 it must reach P_3P_4 in Ω_2, etc. Hence γ' spirals around 0 in the same manner as γ. This proves (20.5).

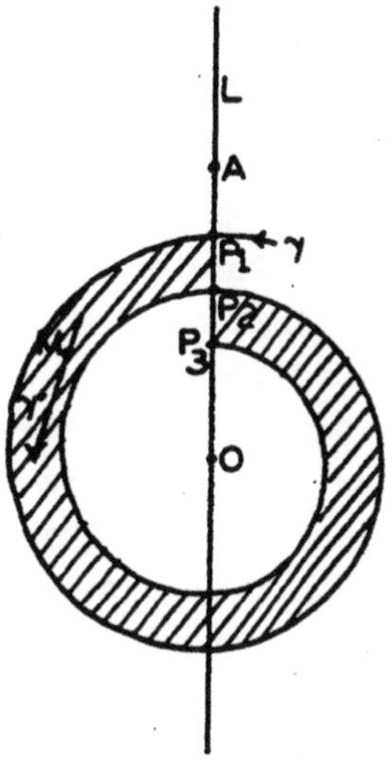

Figure 23

21. (21.1) Continuing with the analysis of the limiting sets let us take any closed characteristic δ and examine its relation to the characteristics issued from the neighboring points. Choose any point P of δ and a segment σ transverse to δ at P and situated in a certain line L. Referring to (II, 14.8), it is known that starting from any point $M \in \sigma$ and following $\gamma^{+}(M)$ we first en-

counter L at a point M_1 and $M \longrightarrow M_1$ defines a topological mapping θ of the segment σ into another segment σ_1 in L. The point P is a fixed point of θ. Further properties will now be brought out.

Since $\gamma^+(M)$ cannot cross δ, the point M_1 is on the same side of δ as M, i. e. on the same side of P as M in L. Since the functions involved in determining M_1 as a function of M are all analytical, M_1 is an analytical function of M. This implies that for σ sufficiently small, either $M = M_1$ for all M or else only for $M = P$. In the first case all the characteristics meeting σ on the side considered are closed; in the second none except δ are closed. Suppose we have the second situation and for one M the point M_1 is between M and P as in fig. 24. Then the same situation will hold for all points between P and M, for otherwise by a well-known continuity argument θ would possess a fixed point between P and M. Under the circumstances, all the characteristics meeting σ on the same side as M will spiral towards δ with increasing t. Let M, M_1, M_2, ... be the successive intersections of any one of them, say γ with σ. By the above M, M_1, M_2, ... are met in that order on σ and they are all on the segment $\overline{MP}$. Hence $\{M_n\}$ has a limit P' on PM. Since P' is a fixed point of θ we must have $P' = P$, and so $\{M_n\} \longrightarrow P$. Since this occurs on any segment transverse to δ, as a point follows $\Lambda^+(M)$ with $t \longrightarrow +\infty$, the point tends to δ. This means that $\Lambda^+(\gamma) = \delta$, and that γ spirals forward around δ towards δ. Moreover all the characteristics issued from points near enough to δ on the same side as

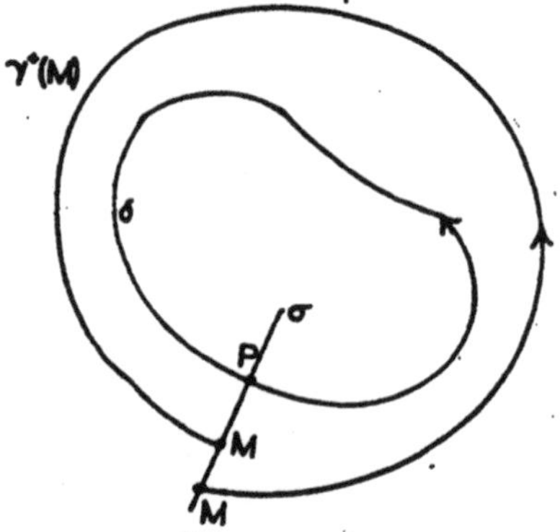

Figure 24

γ behave likewise.

If the point M_1 in the above argument is such that M is between P and M_1, then following $\gamma^-(M_1)$ backwards, we would find that γ and all the characteristics on the same side issued from points near enough to δ spiral towards δ as $t \longrightarrow -\infty$. In this case $\Lambda^-(\gamma) = \delta$.

(21.2) When the characteristics spiral towards [away from] δ on a given side with increasing time, then they do the opposite with decreasing time, and so they are then orbitally stable [unstable] on that side. Notice that the behavior on the two sides of δ need not be alike. If δ is orbitally stable [unstable] on both sides, then it is orbitally stable [unstable] in our usual sense. If it does not behave alike on both sides, then it is orbitally conditionally stable. We shall refer briefly to the three possibilities as _stable, unstable_, and _semi-stable_. That all three actually arise will be shown in the examples later.

(21.3) To sum up, then, when δ is closed the only characteristics γ which have limit points on δ, i. e. whose limit sets meet δ, are δ itself or those which spiral towards or away from δ. In the first case $\Lambda^+(\gamma)$, in the second $\Lambda^-(\gamma)$ is δ itself. The behavior relative to δ is clearly in accordance with (19.1).

22. (22.1) The remaining possibility for $\Lambda^+(\gamma)$ is to consist of critical points and non-closed characteristics. Let δ be one of the latter. Now we may prove by the same arguments as in (20.2) that if δ is a separatrix and $\Lambda^+(\delta)$ contains an ordinary point then δ must be closed. Since the δ here considered is not closed its $\Lambda^+(\delta)$ consists only of critical points, and since $\Lambda^+(\delta)$ is connected it consists of a single critical point, say A. The rest of the argument will be based upon the behavior of γ and δ in the vicinity of A.

We shall first show that σ cannot spiral around A. For γ passes arbitrarily near every point of σ, and hence arbitrarily near A. Hence if σ spirals around A, so does γ (20.5) and therefore $\Lambda^+(\gamma) = A$ (20.1.1) contrary to assumption. Thus σ tends to A in a definite direction (17.5).

(22.2) Let Γ be a circumference of radius ε so small that neither Γ nor its interior contain critical points other than A. Since Γ intersects σ, it is not a characteristic. Hence σ meets Γ a finite number of times, for otherwise both being analytical they would coincide. On the other hand, the number of points of Γ where the characteristic is tangent to Γ is finite. For taking A as the origin, they are the points where

$$r\frac{dr}{dt} = xP + yQ = 0.$$

Since $xP + yQ$ is analytical if it has an infinite number of zeros on Γ it contains Γ. Hence again Γ must be a characteristic which has just been ruled out.

We conclude therefore that if P is the last intersection of σ with Γ, as σ is described with increasing time, there is one arc without contact λ containing P as an interior point. Since P is a limit-point of any $\gamma^+(M)$, γ will intersect λ from a certain point P_1 in a succession of points $P_1, P_2, \ldots$, which tend monotonely to P on λ on one side of P. At these points, and indeed throughout λ the characteristics enter Γ.

(22.3) Let us follow γ from P_n on. Since $\gamma^+(P_n)$ has limit-points exterior to Γ, there is a first crossing Q_n of Γ by γ beyond P_n. Let γ_n be the arc of γ described from P_n to Q_n. The points Q_n are all exterior to λ. Moreover, since the γ_n are disjoint, the sequence $Q_1, Q_2, \ldots$, is monotone on Γ and advances in the opposite direction from $P_1, P_2, \ldots$. Hence $\{Q_n\}$ has a limit Q which is not in λ. We denote by δ_1 the characteristic through Q. Since $\gamma^+(M)$ contains every Q_n for n above a certain value, Q is in $\bar{\gamma}^+(M)$ and hence in $\Lambda^+(\gamma)$. Hence δ_1 is a separatrix.

By the same argument as above for δ we show that δ_1 intersects Γ in at most a finite number of points. Moreover a characteristic rectangle containing Q and with its axis in δ_1 will contain Q_n for n sufficiently high, and hence it will contain an arc of γ_n terminating in Q_n. From this follows that δ_1 followed backwards from Q must enter Γ. Suppose that when δ_1 is thus followed backwards it leaves the interior of Γ and let it occur for the first time at R. Construct a characteristic rectangle whose axis contains the closed arc QR (Figure 25). Since γ_n, for n sufficiently high, has arcs in the rectangle, it is seen by reference to the figure that γ_n must traverse Γ when followed from P_n towards Q_n before Q_n is reached. Since this contradicts the assumption as to Q_n, δ_1 followed backwards never leaves the closed circular region Ω bounded by Γ. Hence $\Lambda^-(\delta_1)$ tends to a critical point in Γ and this point can only be A.

(22.4) To sum up then the true diagram of the characteristics in Γ is in accordance with Figure 26, and is in every way analogous to that of a saddle point.

As we follow δ_1 we shall reach a critical point A_1 (which may be A itself), then leave it with a separatrix δ_2, etc. Since γ must return to one of the points P_n, we shall ultimately return to δ, after having described in the process the full set $\Lambda^+(\gamma)$. The structure of $\Lambda^+(\gamma)$ is thus that of a linear graph. The argument of (21.1) is applicable here to show that γ and any characteristic passing sufficiently near G on the same side as γ all spiral around G in the same direction and it need not be repeated. Let U be the component of the complement of G in the sphere containing γ. The boundary of U contains δ, hence more than two points. Therefore by

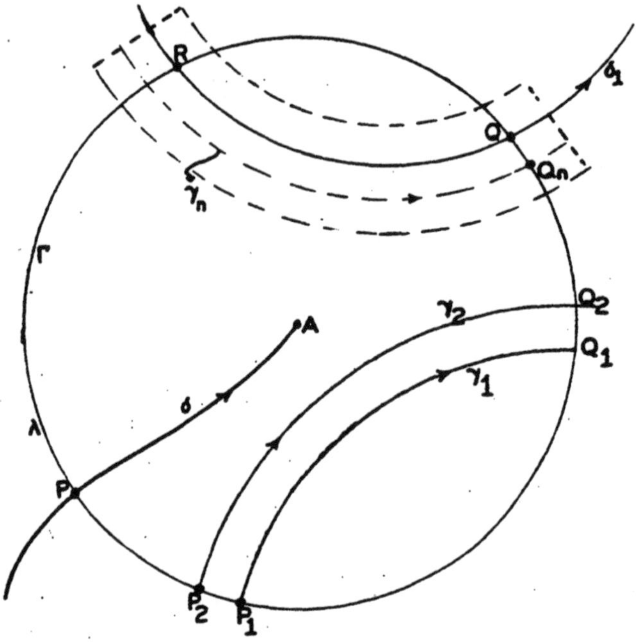

Figure 25

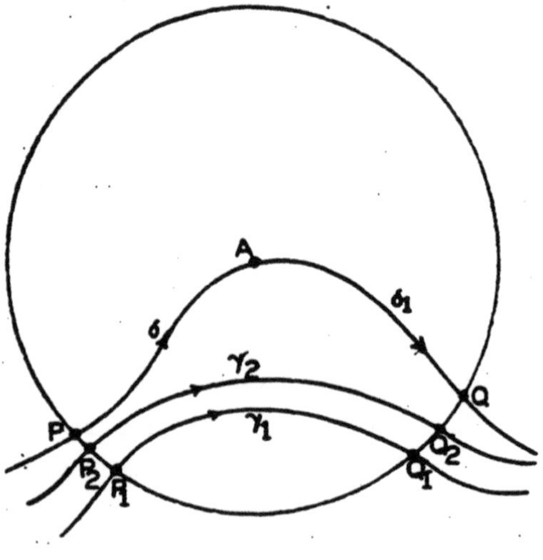

Figure 26

the conformal mapping theorem U is a 2-cell. This completes the proof of theorem (19.1).

23. Some complements on closed characteristics.

(23.1) Suppose that γ, δ are closed characteristics bounding an annular region free from singular points or other closed characteristics. We then say that γ, δ are consecutive.

(23.2) Two consecutive closed characteristics γ, δ cannot both be stable or unstable on the sides facing one another. (Poincaré)

That is to say, if say γ is interior to δ then it is not possible to have γ stable [unstable] outside and δ stable [unstable] inside. Suppose, in fact, the assertion false. Replacing if need be t by -t, we may dispose of the situation so that both γ, δ are unstable in the sides facing one another. Choosing now characteristics γ', δ' in the annular region respectively very near γ, δ and suitable transverse segments, we will have the configuration of Figure 27. Let U be the inner annular region bounded by arcs of γ', δ' together with the segments MM', NN'. Any characteristic starting from a point of the boundary of $\overline{U}$ remains in $\overline{U}$. Since $\overline{U}$ is free from critical points it must contain a new closed characteristic. Since this contradicts the assumption that γ, δ are consecutive (23.2) is proved.

(23.3) Example. Consider the general equation in polar coordinates

(23.3.1) $$\frac{dr}{d\theta} = rf(r^2)$$

where f is a polynomial. Setting $\theta = t$, and passing to rectangular coordinates, we obtain the system

(23.3.2) $$\frac{dx}{dt} = -y + xf(x^2+y^2)$$
$$\frac{dy}{dt} = x + yf(x^2+y^2)$$

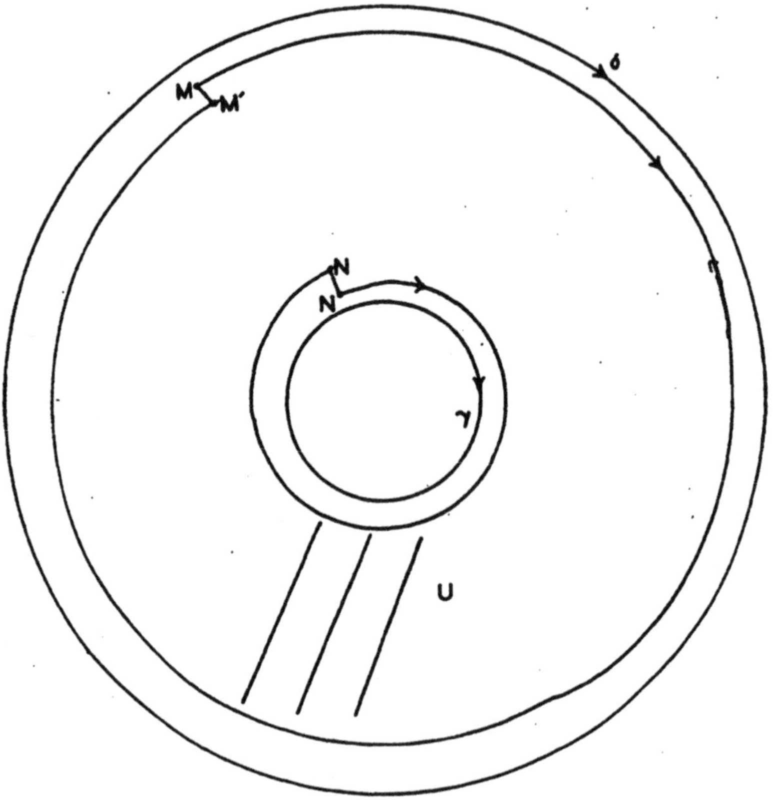

Figure 27

which is of the basic type (1.1). The basic form of the characteristics is more conveniently investigated, however, in the polar form. Since only the positive roots of $f(z)$ matter, let us set $f(z) = z^k g(z)h(z)$ where $g(z) > 0$ for $z > 0$ and $h(z)$ has only positive roots. It is also not a genuine restriction to assume $h(0) > 0$, since this may be achieved by the unimportant change of θ into $-\theta$.

Now if α is a root of $h(z)$ the circle C_α: $r^2 = \alpha$ is a closed characteristic. There are two possibilities:

(a) The root α is of odd order. Then $h(r^2)$ changes its sign as r^2 crosses α. Suppose that it goes from - to +. Since $\frac{dr}{dt}$ changes sign from - to + as r^2 crosses α any half-characteristic near C_α inside the circle will spiral away from C_α, and likewise for any half-characteristic near C_α outside the circle. Hence C_α is unstable (on both sides). If the change of sign of $h(z)$ is in the opposite direction, then C_α is stable.

(b) The root α is of even order. Then $h(r^2)$ does not change sign as r^2 crosses α. If $h(r^2) \geq 0$ near α then along a characteristic near C_α, r^2 must increase and so C_α is stable inside but unstable outside. If $h(r^2) \leq 0$ near α, C_α is stable outside and unstable inside. Thus C is semi-stable in both cases.

If α, β are consecutive roots of $h(z)$, then along any characteristic γ in the region between C_α and C_β the sign of $\frac{dr}{dt}$ is fixed and so r^2 is monotone increasing or decreasing. Hence γ spirals away from one of the two circles and towards the other. Thus the only closed characteristics in the finite plane are the circles C_α. Beyond the last circumference C_α the characteristics spiral to infinity if the leading term of $h(z)$ is positive, away from infinity in the contrary case.

The origin is manifestly a focus. The characteristics passing near the origin spiral away from the origin since $g(0)h(0) > 0$. Thus the origin is an unstable

CHAPTER VI

APPLICATION TO CERTAIN EQUATIONS OF THE SECOND ORDER

§1. EQUATIONS OF THE ELECTRIC CIRCUIT

In this concluding chapter a number of the theories developed so far converge as it were on the applications to certain noteworthy equations of the second order, of which those of the electric circuit are typical. The central problem in most applications is to find the possible periodic motions. Referred, in our customary manner, to the related system of the first order, the periodic motions correspond to closed characteristics. Thus their investigation will constitute our most important problem.

Besides the existence theorems, little more than the following properties will be required:

(1.1) The three types of simple critical points: node, focus, saddle point, their characterization by means of the characteristic roots, and their indices.

(1.2) The sum of the indices of the critical points in the region bounded by a closed characteristic is unity.

(1.3) If a closed characteristic γ is interior to a closed characteristic δ, and there are no other closed characteristic and no critical points between them, then of the two sides of γ, δ facing one another, the one is stable and the other unstable.

2. The differential equation

$$m\frac{d^2x}{dt^2} + l\frac{dx}{dt} + kx = F(t) \tag{2.1}$$

where m, l, k are positive constants arises in a number of applications. Thus it may represent the motion of a particle of mass m held by a "linear" spring with spring constant k, undergoing a resistance (dissipative force) proportional to the velocity and subjected to an exterior disturbance F(t). There is a parallel electrical interpretation, generally written

$$L \frac{d^2 i}{dt^2} + R \frac{di}{dt} + \frac{1}{C} i = \frac{dE(t)}{dt} \tag{2.2}$$

for the so-called L, R, C circuit with impressed voltage E(t) and current i. Under certain circumstances notably in circuits with vacuum tubes, similar phenomena lead to equations of the same form save that the coefficients l, k in (2.1) or R, $\frac{1}{C}$ in (2.2) cease to be constant. The differential equation assumes thus the form

$$\frac{d^2x}{dt^2} + f(x) \frac{dx}{dt} + g(x) = F(t) \tag{2.3}$$

with f even and g odd. Generally the two functions are only known empirically and may merely be given by means of a graph. An important problem arising in practice concerns the existence of possible self-starting, or "autonomous" oscillations, i.e., oscillations arising even in the absence of a disturbing force. The induced or forced oscillations are those which are caused as a response to an oscillatory force, i.e., to a function F(t) which is periodic. The theory of autonomous oscillations reduces to that of a pair of equations of the first order with constant field; that is to say, to the systems discussed in Chapter V. Thus much more is known regarding autonomous than forced oscillations, and so we shall deal primarily with autonomous oscillations.

3. Consider first the particularly simple equation

$$\frac{d^2x}{dt^2} + g(x) = 0. \tag{3.1}$$

Dynamically (3.1) represents the rectilinear motion of a particle of mass one subjected to a force $-g(x)$. For simplicity we will assume $g(x)$ analytic for all x. If we introduce

$$G(x) = \int_0^x g(x)dx, \tag{3.2}$$

then $g(x) = \frac{dG}{dx}$, and so the dynamical system is of the type known as <u>conservative</u>, a term frequently applied to (3.1) itself. Notice that G is the so-called potential energy.

The equation (3.1) may be integrated as follows: Multiply both sides by $\frac{dx}{dt}\,dt$ and integrate, thus obtaining

$$\frac{1}{2}\left(\frac{dx}{dt}\right)^2 + G(x) = h. \tag{3.3}$$

The first term represents the kinetic energy of the particle and (3.3) expresses the law of conservation of energy as applied to the particle.

A further integration yields

$$t - t_0 = \int_{x_0}^{x} \frac{dx}{\sqrt{2(h-G(x))}} \tag{3.4}$$

which is usually the starting point for discussing the solution of the given equation.

To obtain further insight into the situation, it is better, however, to reduce the system to

$$\frac{dx}{dt} = y, \qquad \frac{dy}{dt} = -g(x) \tag{3.5}$$

Then (3.3) yields as the equation of the characteristics

$$\frac{1}{2}y^2 + G(x) = h. \tag{3.6}$$

Consider the curve Γ: $y = G(x)$ and suppose that the part of Γ below $y = h$ includes a finite arc AB with its endpoints A(a,h), B(b,h). Then $y = \pm\sqrt{2(h-G(x))}$ will represent an oval in the strip $a \leq x \leq b$ which is a closed characteristic. We thus have here a continuous family of closed characteristics depending upon the parameter h. We forego a more detailed treatment and will merely make a certain observation regarding the period. For evident reasons of symmetry, the two halves of the oval, above and below the x axis are described in the same time. Hence the period of the resulting oscillation, obtained from (3.4) is:

$$T = 2\int_a^b \frac{dx}{\sqrt{2(h-G(x))}}$$

and it is not difficult to show that it depends generally upon h, and is not constant. When (2.1) is linear, however, it is well known that T is constant.

§2. LIÉNARD'S EQUATION

4. In a noteworthy, but not very well known paper (Étude des oscillations entretennes, Revue Générale de l'electrécité, XXIII (1928), 901-946) the French engineer A. Liénard investigated at length a very general equation with a dissipative middle term

$$\frac{d^2x}{dt^2} + f(x)\frac{dx}{dt} + g(x) = 0 \tag{4.1}$$

to which we shall refer as Liénard's equation. Liénard makes a certain number of assumptions, very largely fulfilled in practice, and described below. The same equation, with still weaker assumptions, has been treated by N. Levinson and O. K. Smith (A general equation for relaxation oscillations, Duke Journal, IX (1942), 382-403) with results similar to those of Liénard, but requiring decidedly more complicated considerations. For the sake of simplicity, we confine our attention to Liénard's work.

(4.2) If we set

$$F(x) = \int_0^x f(x)dx, \qquad G(x) = \int_0^x g(x(dx$$

$$y = \frac{dx}{dt} + F(x), \qquad \lambda(x,y) = \frac{y^2}{2} + G(x)$$

then in the "spring" interpretation $\frac{y^2}{2}$ is the kinetic energy, G(x) the potential energy and (4.1) yields, after multiplication by y:

$$d\lambda = Fdy.$$

The integral $\int Fdy$ taken along a characteristic is the energy <u>dissipated</u> by the system.

Following Liénard we shall give a rather full discussion of (4.1) under the assumption given below:

(4.2.1) f is even, g is odd, $xg(x) > 0$ for all x, $f(0) < 0$;

(4.2.2) f and g are analytic for all x (weaker assumptions could be made but analyticity is convenient);

(4.2.3) $F \longrightarrow +\infty$ with x;

(4.2.4) F has a single positive zero x = a and is monotone increasing for $x \geq a$.

It may be observed that van der Pol's equation

$$(4.3) \qquad \frac{d^2x}{dt^2} - \mu(1-x^2)\frac{dx}{dt} + x = 0, \qquad \mu > 0,$$

so important in vacuum tube circuits, is of the type under consideration.

(4.4) <u>Theorem</u>. <u>Under the assumptions</u> (4.2.1, ..., 4.2.4) <u>equation</u> (4.1) <u>possesses a unique periodic solution which is orbitally stable</u> (Liénard).

In terms of x, y, (4.1) is equivalent to

$$(4.5)\qquad \begin{cases} \dfrac{dx}{dt} = y - F(x) \\ \dfrac{dy}{dt} = -g(x). \end{cases}$$

To prove (4.4), we merely need to prove:

(4.6) *The system (4.5) possesses a single closed characteristic which is orbitally stable.*

Observe the following simple properties:

(4.7.1) *If $x(t)$, $y(t)$ is a solution of (4.5), so is $-x(t)$, $-y(t)$. That is to say, a curve symmetric to a characteristic with respect to the origin is likewise a characteristic.*

(4.7.2) *The only critical point of the system (4.5) in the finite plane is the origin. Hence any closed characteristic must surround the origin.*

(4.7.3) *The slope of a characteristic Γ is given by*

$$\frac{dy}{dx} = \frac{-g(x)}{y-F(x)} .$$

Referring to the form of the curve Δ: $y = F(x)$ (fig. 1), we see that outside the origin on the y axis, the tangents to the characteristics are horizontal while on Δ they are vertical.

Since $xg > 0$, it is a consequence of (4.5) that y decreases along a characteristic Γ to the right of the y axis and increases to the left of the y axis. On the other hand, from (4.5) follows that x increases when Γ is above Δ, and decreases otherwise. Hence Γ has the aspect of figure 1. We denote by α the abscissa of B and write Γ_α for Γ.

(4.7.3) We are now in position to specify certain conditions for Γ_α to be a closed characteristic. At all events it will be necessary that $OA' = OA$, since otherwise Γ_α would have a self-crossing. Hence Γ_α must intersect each axis in two and only two points. It follows that we must have $OA = -OC$. For suppose that $|OA| \neq |OC|$

and let A_1, C_1 be the symmetrical of A, with respect to O. By (4.7.1) the curve symmetric to Γ_α with respect to the origin is a closed characteristic Γ_1 through A_1, C_1. Since the y axis is not tangent to Γ_α, and the segment

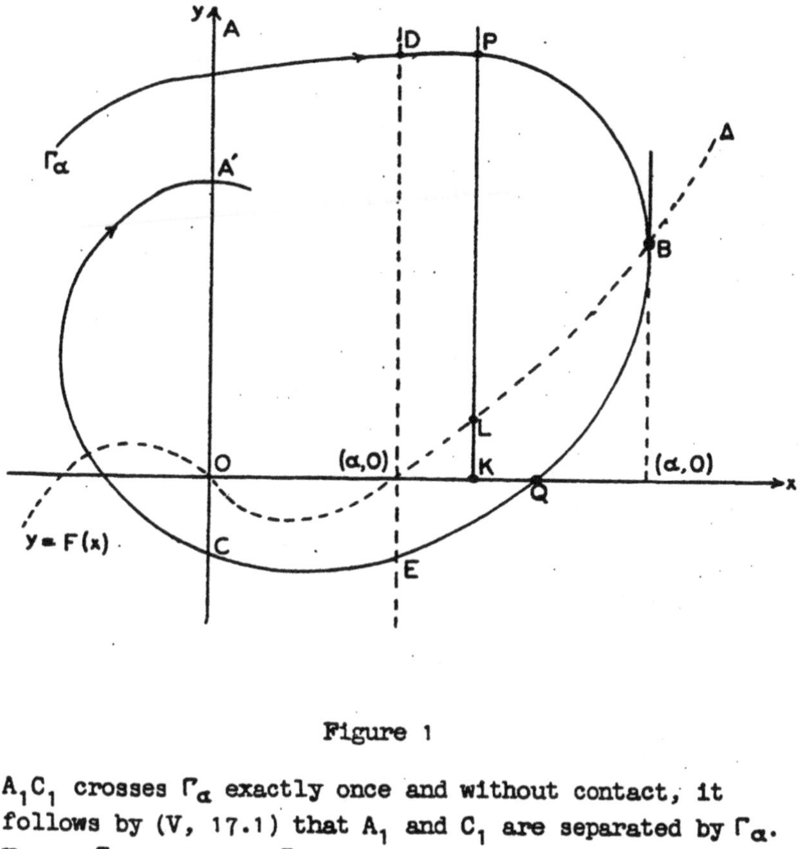

Figure 1

A_1C_1 crosses Γ_α exactly once and without contact, it follows by (V, 17.1) that A_1 and C_1 are separated by Γ_α. Hence Γ_1 must meet Γ_α, which is ruled out, and so $OA = -OC$.

Conversely, suppose $OA = -OC$. The curve symmetric to the arc AC with respect to the origin is an arc of characteristic joining A to C to the left of the y axis, and so together with AC it makes up a closed characteristic. Thus a n. a. s. c. for a closed characteristic is that OA = -OC. Since $\lambda(0,y) = \frac{y^2}{2}$, we may assert:

(4.8) <u>A n. a. s. c. for Γ to be closed is that</u> $\lambda(A) = \lambda(C)$.

5. (5.1) The integrals to be written presently are all curvilinear and along Γ. Consider

$$\varphi(\alpha) = \int_{ABC} d\lambda = \int_{ABC} F dy = \lambda_C - \lambda_A.$$

If $\alpha \leqq a$, both F and dy are < 0 and so $\varphi(\alpha) > 0$, $\lambda_C > \lambda_A$. Hence (4.8) Γ_α cannot be closed.

(5.2) Suppose now that $\alpha \geqq a$, i. e. Γ_α is as in fig. 1. We may now introduce

$$\varphi_1(\alpha) = \int_{AD} d\lambda + \int_{CE} d\lambda, \qquad \varphi_2(\alpha) = \int_{DBE} d\lambda,$$

so that $\varphi(\alpha) = \varphi_1(\alpha) + \varphi_2(\alpha)$. Along AD or CE we may write

$$d\lambda = F \frac{dy}{dx} dx = \frac{(-F)g}{y-F} dx.$$

Since $F < 0$ for $x < a$, $d\lambda$ is positive when Γ is described forward from A to D or from E to C and so $\varphi_1(\alpha) > 0$. On the contrary along DBE we have $d\lambda < 0$ and so $\varphi_2(\alpha) < 0$.

The effect of increasing α is to raise the arc AD and to lower CE, thus increasing $|y|$ for given x. Since for φ_1 the limits of integration are fixed, the result will be a decrease in $\varphi_1(\alpha)$.

Concerning φ_2 we proceed slightly differently. The transformation $X = F(x)$, $Y = y$ is topological to the right of the line DE. The arc DE goes into an arc D^*E^* with end points on the Y axis and $-\varphi_2(\alpha)$ is the area between D^*E^* and this axis. Now an increase in α causes D^* to rise say to D_1^* and E^* to be lowered say to E_1^*. The new arc $D_1^*E_1^*$ does not meet D^*E^* and so the effect is an increase in the area between D^*E^* and the Y axis, i.e. again a decrease in $\varphi_2(\alpha)$.

It follows that when $\alpha \geq a$, $\varphi(\alpha) = \varphi_1(\alpha) + \varphi_2(\alpha)$ is monotone decreasing. Notice that $\varphi(\alpha) = \varphi_1(\alpha) > 0$, for $\alpha \leq a$.

We will now show that $-\varphi_2(\alpha) \to +\infty$ with α. Since $d\lambda < 0$ on DBE we merely need to show that $-\int d\lambda \to +\infty$ along some subarc. Take K in fig. 1 fixed and $\alpha > OK$. We have

$$\int_{PBQ} d\lambda = \int_{PBQ} F(x)dy < -PK \cdot KL.$$

Since PK is arbitrarily large, the integral $\to -\infty$ and so does $\varphi_2(\alpha)$.

Now φ is monotone decreasing from $\varphi(a) > 0$ to $-\infty$ as α ranges from a to $+\infty$. Hence $\varphi(\alpha)$ vanishes once and only once say for $\alpha = \alpha_0$ and so by (4.8) there is one and only one closed characteristic which is Γ_{α_0}.

When $\alpha < \alpha_0$ we have $\varphi(\alpha) > 0$, while $\varphi(\alpha) < 0$ when $\alpha > \alpha_0$. Hence $\lambda_C > \lambda_A$ for $\alpha < \alpha_0$, while $\lambda_C < \lambda_A$ when $\alpha > \alpha_0$. If A_0, C_0 correspond to Γ_{α_0}, we see then that C is nearer to Γ_{α_0} than A when $\alpha < \alpha_0$. Repeating the same argument, which may be done in view of (4.7.3), we find that A' is then nearer to Γ_{α_0} than A. The same thing will clearly hold for the same reasons when $\alpha > \alpha_0$. This is what fig. 1 shows. Hence Γ_{α_0} is orbitally stable. Thus Liénard's theorem is proved.

6. What is the effect of omitting condition (4.2.4), i.e. of assuming that F ceases to be monotone? The general aspect of Γ will then be as in fig. 2. It is not difficult to see that we shall still have $\varphi(\alpha) > 0$ for α small and < 0 for α large. Hence it will have at least one zero, but may very well have several $\alpha_1, \alpha_2, \ldots$ thus giving rise to a succession of "concentric" closed characteristics $\Gamma_1, \Gamma_2, \ldots$, and Γ_i will be orbitally stable [unstable] whenever $\varphi'(\alpha_i) < 0$ [> 0]. This is proved as at the end of (5). It is readily found that

$$\varphi'(\alpha) = \int_{ACB} f(x)dy$$

and thus $\varphi'(\alpha)$ may be calculated directly without passing through the expression for $\varphi(\alpha)$.

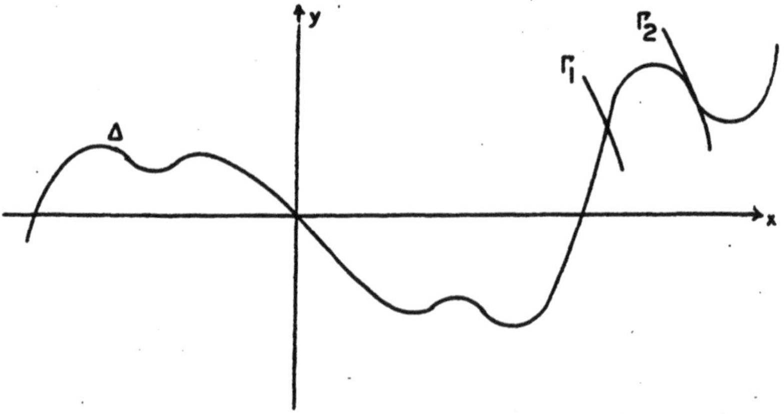

Figure 2

§3. APPLICATION OF POINCARÉ'S METHOD OF SMALL PARAMETERS

7. A large class of equations of the second order is of the type

$$\frac{d^2x}{dt^2} + \omega^2 x = \mu f(x, \frac{dx}{dt}). \tag{7.1}$$

For $\mu = 0$, this represents a harmonic oscillation with a fixed period $\frac{2\pi}{\omega}$. It is to be expected that when f behaves reasonably well and μ is small, there will still occur periodic solutions whose periods differ little from $\frac{2\pi}{\omega}$ as $\mu \to 0$. We shall see that this is effectively the case. To simplify matters, we shall impose the following restrictions generally fulfilled in the applications:

(7.2) $f(x,y)$ <u>is a polynomial</u>;

(7.3) <u>the expansion of</u> f <u>about the origin is of the form</u> $f = ax + by + \dots$, <u>with</u> $b \neq 0$.

(7.4) <u>the polynomial</u> $\varphi(x) = \frac{f(x,0)}{x}$ <u>has a fixed sign or is identically zero</u>.

(7.4.1) Since we may replace μ by $-\mu$ as a parameter, we shall restrict μ to non-negative values and suppose $\varphi(x) \leq 0$.

(7.5) To simplify matters, it is convenient to replace the variable t by $\frac{t}{\omega}$, thus reducing (7.1) to the form

$$(7.6) \qquad \frac{d^2x}{dt^2} + x = \mu f(x, \frac{dx}{dt}),$$

and this will be our basic differential equation. Thus the harmonic period will be 2π.

(7.7) Physically speaking, the systems under consideration represent oscillatory systems with dissipation. In particular, if μ is chosen of the same sign as b then for x and $\frac{dx}{dt}$ both small the dissipation is "negative", i.e. the system absorbs energy. This situation generally implies the presence of an autonomous oscillation, a property amply confirmed in the sequel.

(7.8) Van der Pol's equation is of the type here considered with $f = (1-x^2)y$. Here $b = 1 > 0$ and so by the remark just made, an autonomous oscillation is to be expected. In fact, its existence has already been established in connection with Liénard's equation.

(7.9) The general problem under consideration has been frequently discussed in recent years by many Soviet mathematicians and mathematical physicists. See notably the work of Kryloff-Bogoliuboff, Andronoff-Chajkin: Theory of Oscillations (Russian) and the David Taylor Model Basin Report by N. Minorsky. Our treatment rests on Poincaré's expansion theorem (II,9.1) and is closely related to the treatment of Andronoff-Chajkin.

8. As a first step (7.6) must be replaced in the customary way by the pair of equations of the first order

$$(8.1) \qquad \begin{cases} \frac{dx}{dt} = y \\ \frac{dy}{dt} = \mu f(x,y) - x. \end{cases}$$

The singular points are the solutions of

$$y = 0, \qquad \mu f(x,y) - x = 0,$$

i. e. the points on the x axis where $x(\frac{f(x,0)}{x} - 1)$ vanishes. In view of (7.4) and (7.4.1) the only critical point is the origin.

The characteristic roots s_1, s_2 corresponding to the origin are the roots of

$$(8.2) \qquad \begin{vmatrix} -s, & 1 \\ a\mu - 1, & b\mu - s \end{vmatrix} = s^2 - b\mu s - a\mu + 1 = 0.$$

When μ is small positive s_1 and s_2 are both complex with real part $\frac{1}{2} b\mu$. Hence the origin is asymptotically stable [unstable] when $b < 0$ [$b > 0$]. Therefore there is no closed characteristic arbitrarily near the origin.

9. We shall now investigate (8.1) for the presence of closed characteristics for μ small and positive. The general solution of (8.1) for $\mu = 0$, i. e. of the system

$$(9.1) \qquad \frac{dx}{dt} = y, \qquad \frac{dy}{dt} = -x$$

is given by

$$(9.2) \qquad x = \alpha \sin (t-\theta), \qquad y = \alpha \cos (t-\theta)$$

where α, θ are arbitrary constants. The corresponding characteristic is the circle

$$(9.3) \qquad \Delta_\alpha : x^2 + y^2 = \alpha^2.$$

We are particularly interested in finding the true circles Δ_α (i.e. with $\alpha \neq 0$) which are limits as $\mu \to 0$ of closed characteristics of (8.1). Such a circle will be called a circle generating closed characteristics. The corresponding solution $x = \alpha \sin (t-\theta)$ of (7.6) will be called a harmonic generating periodic solutions.

Choosing the particular solution $x = \alpha \sin t$, $y = \alpha \cos t$, $\alpha > 0$, is equivalent to assuming that t is counted from the time when Δ_α crosses the positive y axis at $(0,\alpha)$. This assumption will be made throughout.

If we consider the characteristics of (8.1) as solutions of the system obtained by adjoining $\frac{d\mu}{dt} = 0$ to (8.1) and apply (II, 14.3) to this system, we find that all the characteristics passing near $(0,\alpha)$ for μ small will intersect the y axis in a point $(0,\lambda)$ where λ is very near α and hence in particular $\lambda > 0$. Hence for a time exceeding 2π, and by (II, 9.1), we find that the characteristics passing through $(0,\lambda)$, with $(\lambda-\alpha)$ sufficiently small, will be represented by a system

$$(9.4)\quad \begin{cases} x = \alpha \sin t + A_0(\lambda,t) + \mu A_1(\lambda,t) + \mu^2 A_2(\lambda,t) + \dots \\ y = \alpha \cos t + A_0'(\lambda,t) + \mu A_1'(\lambda,t) + \mu^2 A_2'(\lambda,t) + \dots \end{cases}$$

where the A's are holomorphic in (t,λ) for $-\varepsilon < t < 2\pi+\varepsilon$, $|\lambda-\alpha| < \eta$, the accents denote derivation as to t, and the representation (9.4) is valid in $|\mu| < \rho$ for suitably small positive numbers ε, η, ρ.

For $\mu = 0$ the solution (9.4) reduces to the solution of the system (9.1) such that $x = 0$, $y = \lambda$ for $t = 0$, i.e. it reduces to $x = \lambda \sin t$, $y = \lambda \cos t$. Therefore $A_0(\lambda,t) = (\lambda-\alpha) \sin t$ and so (9.4) may be written

$$(9.5)\quad \begin{cases} x = \lambda \sin t + \mu A_1(\lambda,t) + \mu^2 A_2(\lambda,t) + \dots \\ y = \lambda \cos t + \mu A_1'(\lambda,t) + \mu^2 A_2'(\lambda,t) + \dots \, . \end{cases}$$

For $t = 0$, we must have $x = 0$, $y = \lambda$ whatever μ and this implies

$$(9.6)_n \qquad A_n(\lambda,0) = A_n'(\lambda,0) = 0.$$

If we substitute x,y from (9.5) in μf, and apply (III, 9.2) we obtain for the solution of (8.1) the relations

$$(9.6)\quad \begin{cases} x = \lambda \sin t + \mu \int_0^t f(x,y) \cdot \sin (t-u)dt \\ y = \lambda \cos t + \mu \int_0^t f(x,y) \cdot \cos (t-u)dt \end{cases}$$

where under the integration signs x,y are the solutions expressed as functions of u.

We must now endeavor, following Poincaré (IV, 18), to specialize the solution (9.6) to one with a period $2\pi + \tau(\mu)$, where $\tau(\mu) \to 0$ with μ, and if possible is an analytic function of μ. Expressing the fact that (9.6) has the period in question we obtain the basic relations:

$$(9.7)\quad H(\lambda,\mu,\tau) = \lambda \sin \tau + \mu \int_0^{2\pi+\tau} f(x,y) \sin(t-u)du = 0$$

$$(9.8)\quad K(\lambda,\mu,\tau) = \lambda (\cos \tau - 1) + \mu \int_0^{2\pi+\tau} f(x,y)\cos(\tau-u)du = 0.$$

Since we are investigating the existence of a solution of (7.6) tending, when $\mu \to 0$, to the circle (9.3), we assume $\lambda - \alpha$ small, and of course $\alpha \neq 0$. It is readily found that the Jacobian

$$\left(\frac{\partial(H,K)}{\partial(\lambda,\tau)}\right)_{(\alpha,0,0)} = 0$$

and so we cannot apply the implicit function theorem for the two variables λ,τ to our system. However

$$H(\alpha,0,0) = 0, \qquad \left(\frac{\partial H}{\partial \tau}\right)_{(\alpha,0,0)} \neq 0$$

and so (9.7) may be solved for τ as a holomorphic function of λ, μ in the neighborhood of $(\alpha,0)$ and we will have $\tau(\lambda,0) = 0$. Thus

$$(9.9)\qquad \tau(\lambda,\mu) = B_1(\lambda)\mu + B_2(\lambda)\mu^2 + \dots$$

where the $B_i(\lambda)$ are holomorphic in a certain circular region $|\lambda-\alpha| < \xi$.

It will be convenient to have the value of $B_1(\lambda)$ later. We find at once

$$B_1(\lambda) = \left(\frac{\partial\tau(\lambda,\mu)}{\partial\mu}\right)_{(\lambda,0)} .$$

Now by differentiating (9.7) as to μ with $\tau = \tau(\lambda,\mu)$ there comes:

$$\frac{\partial H}{\partial\mu} = \lambda \cos\tau\cdot\frac{\partial\tau}{\partial\mu} + \int_0^{2\pi+\tau} f(x,y)\sin(\tau-u)du + \mu\,[\qquad] = 0,$$

where the value of the square bracket is omitted since it is not needed. Since $\tau(\lambda,0) = 0$, we have

$$\left(\frac{\partial\tau}{\partial\mu}\right)_{(\lambda,0)} = -\int_0^{2\pi} f(\lambda \sin u, \cos u)\sin u\, du.$$

Hence

(9.10) $$B_1(\lambda) = \frac{1}{\lambda}\int_0^{2\pi} f(\lambda \sin u, \lambda \cos u)\sin u\, du.$$

10. Once $\tau(\lambda,\mu)$ is known it is substituted in (9.8) which is then replaced by

(10.1) $$K(\lambda,\mu,\tau(\lambda,\mu)) = 0.$$

The left-hand side is a series in $(\lambda-\alpha)$ and μ containing no term independent of μ. Thus

$$K(\lambda,\mu,\tau(\lambda,\mu)) = \mu K_1(\lambda,\mu),$$

and the solution $\lambda(\mu)$ of (10.1) such that $\lambda(0) = \alpha$, will satisfy

(10.2) $$K_1(\lambda,\mu) = 0.$$

In order that (10.2) possess a solution of this nature we must have

(10.3) $$K_1(\alpha,0) = 0.$$

Now

$$K_1(\alpha,0) = \left[\frac{\partial K(\lambda,\mu,\tau(\lambda,\mu))}{\partial\mu}\right]_{(\alpha,0)}$$

$$= \int_0^{2\pi} f(\alpha \sin u, \alpha \cos u) \cos u \, du,$$

and so we must have

$$(10.4) \quad \Phi(\alpha) = \int_0^{2\pi} f(\alpha \sin u, \alpha \cos u) \cos u \, du = 0.$$

Since f is a polynomial so is Φ. Thus α must be a real root of a certain algebraic equation.

(10.5) Notice that if α satisfies (10.4) so does $-\alpha$. Therefore $\Phi(\alpha) = \Psi(\alpha^2)$, where Ψ is a polynomial. Thus if Φ has any real root α it also has the positive root $|\alpha|$. It will therefore be sufficient to confine our attention in the sequel to the positive roots of $\Phi(\alpha)$.

11. Choosing now a definite positive root α of Φ, we may expand $K_1(\lambda,\mu)$ as a power series in $\lambda-\alpha$ and μ and we will have

$$(11.1) \qquad K_1(\lambda,\mu) = A(\alpha)(\lambda-\alpha) + B(\alpha)\mu + \dots ,$$

and to discuss stability we require information about $A(\alpha)$. We have

$$(11.2) \quad A(\alpha) = \left(\frac{\partial K_1}{\partial\lambda}\right)_{(\alpha,0)} = \frac{1}{\mu}\left(\frac{\partial K(\lambda,\mu,\tau(\lambda,\mu))}{\partial\lambda}\right)_{(\alpha,0)}.$$

Now

$$\frac{\partial K(\lambda,\mu,\tau(\lambda,\mu))}{\partial\lambda} = \frac{\partial K(\lambda,\mu,\tau)}{\partial\lambda} + \frac{\partial K(\lambda,\mu,\tau)}{\partial\tau}\frac{\partial\tau}{\partial\lambda}$$

$$= \cos\tau - 1 - \sin\tau \cdot \frac{\partial\tau}{\partial\lambda} + \mu\int_0^{2\pi+\tau} \left\{\left[\frac{\partial f}{\partial x}\left(\frac{\partial x}{\partial\lambda} + \frac{\partial x}{\partial\tau}\frac{\partial\tau}{\partial\lambda}\right)\right.\right.$$

$$\left.\left. + \frac{\partial f}{\partial y}\left(\frac{\partial y}{\partial\lambda} + \frac{\partial y}{\partial\tau}\frac{\partial\tau}{\partial\lambda}\right)\right]\cos(\tau-u) - f(x,y,\mu)\sin(\tau-u)\frac{\partial\tau}{\partial\lambda}\right\} du$$

$$+ \mu[f(x,y,\mu)]_{(t = 2\pi+\tau)} \cdot \frac{\partial\tau}{\partial\lambda} .$$

Hence from (11.2) and since by (9.9) $\frac{\partial\tau}{\partial\lambda} = 0$ for $\mu = 0$:

$$A(\alpha) = \int_0^{2\pi} \left(\frac{\partial f}{\partial x}\frac{\partial x}{\partial \alpha} + \frac{\partial f}{\partial y}\frac{\partial y}{\partial \alpha}\right) \cos u \, du,$$

where under the integration sign $f = f(x,y)$, and $x = \alpha \sin u$, $y = \alpha \cos u$. Therefore finally

$$A(\alpha) = \Phi'(\alpha). \tag{11.3}$$

(11.4) Consider now any simple positive root α of Φ, that is to say we suppose that (10.4) holds and that

$$\Phi'(\alpha) \neq 0, \tag{11.5}$$

and we have the corresponding development (11.1). Under the circumstances in view of (11.5), the implicit function theorem asserts that (10.2) has a unique (see 11.6) analytic solution $\lambda(\mu)$ in the neighborhood of $\mu = 0$, such that $\lambda(0) = \alpha$. Substituting then in (9.9) we obtain a similar series $\tau(\mu)$ such that $\tau(0) = 0$. The two functions $\lambda(\mu)$, $\tau(\mu)$ verify (9.7), (9.8) and hence the corresponding solution $(x(\lambda(\mu),\mu,t)), y(\lambda(\mu),\mu,t))$ represents a closed characteristic of (8.1) which tends to the circle Δ_α when $\mu \to 0$. We may also say that $x(\lambda(\mu),\mu,t)$ represents the oscillatory solution of (7.6) which tends to $\alpha \sin t$ when $\mu \to 0$. In other words under the circumstances there is a unique periodic solution of the desired type.

(11.6) We have tacitly assumed the solutions to be real. It is only necessary to observe that the determination of the coefficients of the series involved never makes an appeal to any irrational operations. Hence the coefficients are all real and so are the series.

(11.7) There remains to discuss the question of stability. Let $\gamma_{\lambda,\mu}$ be the characteristic of (8.1) passing through $(0,\lambda)$ and let us assume $\lambda - \alpha$ and μ small. Under the circumstances the solution $\gamma_{\lambda,\mu}$ will cross the positive axis at a point $(0,\lambda_1)$ at a time $2\pi + \tau(\lambda,\mu)$, where $\tau(\lambda,\mu)$ is the solution (9.9) of $H(\lambda,\mu,\tau) = 0$. Let

$$\delta(\lambda,\mu) = \lambda_1 - \lambda. \tag{11.7.1}$$

We have at once

$$\delta(\lambda,\mu) = K(\lambda,\mu,\tau(\lambda,\mu)) = \mu K_1(\lambda,\mu) = (\lambda-\alpha)P(\lambda) + \mu Q(\lambda,\mu),$$

Where P,Q are holomorphic in the neighborhood of $(\alpha,0)$ and $P(\alpha) = A(\alpha) = \Phi'(\alpha) \neq 0$.

We are interested in the behavior of $\gamma_{\lambda,\mu}$ relative to the closed characteristic when the latter is very near Δ_α, i.e. when μ is arbitrarily small. At all events let us choose μ which is assumed positive so that $0 < \mu < (\lambda-\alpha)^2$. Then for $(\lambda-\alpha)$ sufficiently small $\delta(\lambda,\mu)$ has the sign of $(\lambda-\alpha)\Phi'(\alpha)$. Therefore when $\Phi'(\alpha) > 0$, δ is positive for $\lambda > \alpha$ and negative for $\lambda < \alpha$, while for $\Phi'(\alpha) < 0$ the reverse takes place. Thus when $\Phi'(\alpha) > 0$ if $\gamma_{\lambda,\mu}$ starts from a point M outside Δ_α on the positive y-axis, it will intersect the latter at a point M' above M (fig. 3). On the other hand if $\gamma_{\lambda,\mu}$ starts from M inside Δ_α on the positive y-axis, it will intersect the latter at a point M' below M. Hence $\Phi'(\alpha) > 0$ implies asymptotic instability, and clearly when $\Phi'(\alpha) < 0$ we have asymptotic stability.

To sum up we have proved:

(11.8) *Theorem. The generating circles* Δ_α: $x^2 + y^2 = \alpha^2$ *of the closed characteristics of* (8.1) *correspond to the real roots of the polymomial* Φ. *If* α *is such a root and* $\Phi'(\alpha) \neq 0$, *then there is one and only one closed characteristic* Γ_μ *tending to* Δ_α *when* $\mu \to 0$, *and it is stable or unstable accordingly as* $\Phi'(\alpha)$ *is negative or positive. Correspondingly there exists a unique oscillatory solution of* (7.6) (*to within an arbitrary phase* θ) *which tends to the harmonic oscillation* $\alpha \sin(t+\theta)$ *when* $\mu \to 0$, *and whose period* $2\pi + \tau(\mu)$ *tends at the same time to* 2π.

12. *Application to van der Pol's* equation. Here $f = (1-x^2)y$ and so

$$(12.1)\quad \Phi(\alpha) = \int_0^{2\pi} (1-\alpha^2\sin^2 u)\, \alpha \cos^2 u\, du = \pi\alpha\left(1 - \frac{\alpha^2}{4}\right),$$

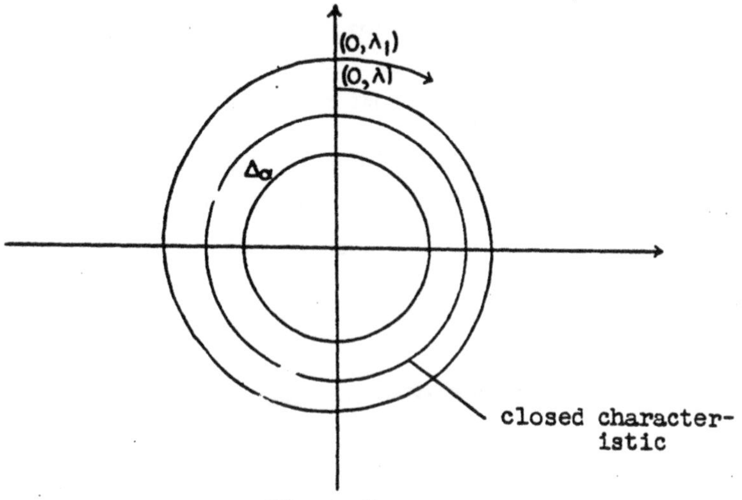

Figure 3

(12.2) $$\Phi'(\alpha) = \pi(1 - \frac{3\alpha^2}{4}) \; . .$$

Hence there is just one generating circle $\Delta_2(\alpha = 2)$. It is the limit of a unique closed characteristic Γ_μ as $\mu \to 0$, and since $\Phi'(2) < 0$, Γ_μ is stable. This is in agreement with the results obtained by Liénard's method, with the important addition that we have actually obtained the generating circle.

(12.3) Another noteworthy result lies near at hand. Referring to (9.9), let us calculate $B_1(\lambda)$. We have here from (9.10)

$$B_1(\lambda) = \int_0^{2\pi} (1-\lambda^2\sin^2 u) \cos u \sin u \, du = 0$$

and therefore

$$\tau(\lambda,\mu) = \mu^2 \{B_2(\lambda) + B_3(\lambda)\mu + \dots \}.$$

In other words for small μ and λ near α, τ is of the order of magnitude of μ^2 or higher. If one neglects everywhere terms of order higher than μ then one may say that τ is zero. That is to say with an approximation sufficient for practical purposes, the non-linear oscillation has the same period 2π as the limiting harmonic oscillation. This is a well known result due to van der Pol.

§4. EXISTENCE OF PERIODIC SOLUTIONS FOR CERTAIN DIFFERENTIAL EQUATIONS

13. In the present section (reprinted from Proc. Nat. Acad., 29 (1943), pp. 1-4) we prove the existence of periodic solutions for the equation

$$\frac{d^2x}{dt^2} + g'(x)\frac{dx}{dt} + f(x) = e(t), \tag{13.1}$$

where $e(t)$ has the period T, and g, f are restricted as stated below. As we shall see, the proof is essentially elementary. The type of equation under discussion generalizes the equation for the response of an electrical series circuit with resistance R, capacity C (both constant) and an inductor with current-flux saturation curve $i = h(\varphi)$. Here (13.1) is the differential equation of the flux with

$$f = \frac{h}{C}, g(x) = Rh.$$

The function $h(x)$ may be satisfactorily represented by an odd polynomial such that $xh(x) > 0$. This particular case suffices to indicate the importance of the periodic solutions of (13.1).

(13.2) Theorem. The following are sufficient conditions in order that (13.1) possess a periodic solution of period T:

I. The derivatives $e'(t)$, $f'(x)$, $g'(x)$ exist for all values of their variables.

II. $e(t)$ has the period T.

III. $\frac{f(x)}{x} \to +\infty$ with $|x|$.

IV. There exist $b, B > 0$ such that

$$|g(x) - bf(x)| \leq B\,|x|.$$

If we set for convenience

$$\frac{f(x)}{x} = F(x), \quad \frac{g(x)}{x} = G(x),$$

then III, IV may be replaced by:

III'. $F(x) \to +\infty$ with $|x|$. (Hence, by IV, $G(x)$ has the same property.)

IV'. $|G(x) - bF(x)| \leq B$.

Remarks. (13.3) Condition I is merely designed to guarantee the existence and uniqueness of a solution with any given initial values of x and $\frac{dx}{dt}$. It would be possible to replace I by a weaker but less easily described condition.

(13.4) The restrictions I, ..., IV are manifestly fulfilled by the equation of the series circuit, but not, for instance, by van der Pol's equation, or generally by the relaxation equation with or without periodic disturbance.

(13.5) Chevalley has proved the existence of periodic solutions for equations (13.1) with f,g still less restricted than ours. His argument rests likewise upon an analogue of Lemma (13.8) applied to the phase space (x, y) with ellipses replaced by the curves

$$2u = y^2 + 2k(x), \quad k = \int_0^x f(x)dx.$$

However we understand that the proof based on elliptic ovals given below is far simpler than his.

(13.6) Method. If we set

$$\frac{dx}{dt} + g(x) = y, \tag{13.6.1}$$

then the solution of (13.1) is equivalent to that of the system consisting of (13.6.1) and of

$$\frac{dy}{dt} + f(x) = e(t), \tag{13.6.2}$$

and our theorem will follow if we can show that the system possesses a solution (x,y) periodic and of period T.

Consider now any point P(x,y). According to the existence theorem (II, 4.1), there is a unique solution or trajectory $\Gamma(x(t), y(t))$ such that $x(0) = x_0$, $y(0) = y_0$. If Q is the point (x(T), t(T)) then $P \to Q$ defines a mapping S of the plane xy into itself and Theorem (13.2) will follow if we can prove that S has a fixed point. For, owing to the periodicity of e(t), the trajectory Γ is likewise uniquely defined by $x(T) = x_0$, $y(T) = y_0$. Hence if Γ returns to P at time T, it is necessarily periodic.

By Brouwer's fixed point theorem then our theorem will follow from

(13.7) Lemma. <u>There is a closed two-cell mapped into itself by S.</u>

We will prove in fact the more precise

(13.8) Lemma. <u>There is a region bounded by an ellipse which is mapped into itself by S.</u>

14. (14.1) Consider the definite quadratic form

(14.1.1) $$2u = ax^2 - 2xy + by^2, \; ab > 1, \; a > 0.$$

Along a trajectory we have

(14.1.2) $$u' = \frac{du}{dt} = (ax-y)(y-xG) + (by-x)(e-xF).$$

Let r, θ be polar coordinates for the plane x,y. We first show that Lemma (13.8) is a consequence of

(14.3) Lemma. <u>When r is above a certain value then</u> $u' < 0$.

In fact assume Lemma (14.3) to hold for $r > r_0$. Since the distance of the ellipse (14.1.1) from the origin $\to +\infty$ with u, we may choose u such that it exceeds r_0, and then Lemma (13.8) will follow from Lemma (14.3). Thus everything reduces to proving the latter. The proof will be divided into two parts.

(a) We first take an α between 0 and $\frac{\pi}{2}$ and assume $\left|\theta - \frac{\pi}{2}\right|$ or $\left|\theta - \frac{3\pi}{2}\right| \leq \alpha$. If $|x|$ is sufficiently large both G and F are positive, and hence in view of $ab > 1$, for $|x|$ large enough

$$aG - F > \frac{1}{b}(G-bF) > -\frac{B}{b}.$$

Hence there is a positive C such that whatever x we have

$$aG - F > -C$$

and therefore

$$\frac{u'}{r^2} < \{C \cos^2 \theta + (a+B) \, |\sin \theta| \, |\cos \theta| - \sin^2 \theta\} - \\ - \frac{e}{r} (b \sin \theta - \cos \theta).$$

The bracket is a continuous function of θ whose value is -1 for $\theta = \frac{\pi}{2} + k\pi$. Therefore for $|\cos \theta|$ sufficiently small, i.e., choosing α sufficiently small, the bracket will be as near -1 as we please. Since e is bounded we may choose r_1 such that for $r > r_1$ the term $\frac{-e}{r}(b \cos \theta - \sin \theta)$ is arbitrarily small in absolute value. Hence we may choose α, r_1 such that $u' < 0$ under the conditions considered.

(b) The point (x,y) is such that $\left|\theta - \frac{\pi}{2}\right|$, and $\left|\theta - \frac{3\pi}{2}\right| > \alpha$, where α is the angle just selected. Evidently

$$\epsilon = \frac{e(b \sin \theta - \cos \theta)}{r \cos^2 \theta}$$

is bounded under the condition here considered. Now

$$(14.4) \quad \frac{u'}{r^2} = -(aG-F+\epsilon) \cos^2\theta + (a+G-bF) \sin \theta \cos \theta - \\ - \sin^2 \theta$$

is a quadratic form in $\sin\theta$, $\cos\theta$ whose discriminant

$$4(aG-F+\epsilon) - (a+G-bF)^2 > 4(ab-1)F + \text{const.} \to +\infty \text{ with } |x|.$$

Therefore the form in (14.4) is definite negative, and so $u' < 0$ if θ is as stated, and if $r > r_2$, where r_2 is chosen large enough.

We conclude then that Lemma (14.3) holds for $r > \max.(r_1, r_2)$ and so Theorem (13.2) is proved.

15. Generalization. Elliptic or other algebraic ovals may be utilized in many other cases to prove the existence of periodic solutions. Thus consider the system

$$(15.1) \qquad \frac{dx_i}{dt} = X_i(x_1,\ldots,x_n,t) + Y_i(x_1,\ldots,x_n)$$

where X_i, Y_i are polynomials in the x_j with degree $X_i < M =$ degree Y_i and the coefficients of X_i are bounded periodic functions of period T. If there exists a positive definite quadratic form

$$2u = \sum a_{ij}x_ix_j$$

such that along a trajectory

$$u' = \sum a_{ij}x_iY_j < 0$$

for $r^2 = \sum x_i^2$ sufficiently large, then the system (15.1) possesses a periodic solution.

(15.2) Let us observe finally that instead of $u' < 0$ for r sufficiently large we could merely ask that its sign be constant. For if u increases with t, it decreases with $-t$, and since the transformation $t \to -t$ does not affect periodic solutions, u increasing with t would be acceptable throughout.

(15.3) The reader may find it profitable to consult the following additional references:

N. Levinson. Existence of periodic solutions for

second order differential equations with a forcing term. Jou. of Math. and Physics, v. 22 (1943), pp. 41-48.

N. Levinson. Transformation theory of non-linear differential equations of the second order. Annals of Math. v. 45 (1944), pp. 723-737.

INDEX

Lightning Source UK Ltd.
Milton Keynes UK
UKHW012319100322
399878UK00001B/44